T0383061

ADDITIVE MANUFACTURING FOR PLASTIC RECYCLING

Sustainable Manufacturing Technologies: Additive, Subtractive, and Hybrid

Series Editors: Chander Prakash, Sunpreet Singh, Seeram Ramakrishna, and Linda Yongling Wu

This book series offers the reader comprehensive insights of recent research breakthroughs in additive, subtractive, and hybrid technologies while emphasizing their sustainability aspects. Sustainability has become an integral part of all manufacturing enterprises to provide various techno-social pathways toward developing environmental-friendly manufacturing practices. It has also been found that numerous manufacturing firms are still reluctant to upgrade their conventional practices to sophisticated sustainable approaches. Therefore, this new book series is aimed to provide a globalized platform to share innovative manufacturing mythologies and technologies. The books will encourage the eminent issues of the conventional and non-conventual manufacturing technologies and cover recent innovations.

Advances in Manufacturing Technology
Computational Materials Processing and Characterization
Edited by Rupinder Singh, Sukhdeep Singh Dhami, and B. S. Pabla

Additive Manufacturing for Plastic Recycling
Efforts in Boosting a Circular Economy
Edited by Rupinder Singh and Ranvijay Kumar

For more information on this series, please visit: https://www.routledge.com/Sustainable-Manufacturing-Technologies-Additive-Subtractive-and-Hybrid/book-series/CRCSMTASH

ADDITIVE MANUFACTURING FOR PLASTIC RECYCLING

Efforts in Boosting a Circular Economy

Edited by
Rupinder Singh and Ranvijay Kumar

CRC Press is an imprint of the
Taylor & Francis Group, an **informa** business

First edition published 2022
by CRC Press
6000 Broken Sound Parkway NW, Suite 300, Boca Raton, FL 33487-2742

and by CRC Press
4 Park Square, Milton Park, Abingdon, Oxon, OX14 4RN

CRC Press is an imprint of Taylor & Francis Group, LLC

ISBN: 978-1-032-02609-1 (hbk)
ISBN: 978-1-032-02610-7 (pbk)
ISBN: 978-1-003-18416-4 (ebk)

DOI: 10.1201/9781003184164

Typeset in Times
by MPS Limited, Dehradun

Contents

Preface

Additive Manufacturing for Plastic Recycling: Efforts in Boosting a Circular Economy presents a comprehensive review of the most recent breakthroughs in the recycling of plastic materials. This book delivers optimized and fool-proof methodology of using additive manufacturing of thermoplastic and thermosetting polymers, keeping in view the concept of recycling (for academic researchers as well as field practitioners). This book highlights commercially established and hybrid thermosetting and thermoplastic processing methods (mechanical as well as chemical), and provides theoretical and practical ideas of combining different processing methods (primary and secondary) for recycling and processing of thermoplastic and thermosetting to boost a circular economy. Also, tertiary recycling of plastic waste followed by economic and environmental justification has been given. Mechanical twin screw extrusion for processing thermoplastics followed by case studies for developing hybrid composite structures for biomedical applications and structural applications (especially for heritage buildings) have been highlighted. Overall, this book is a first-hand source of information for academic scholars and field engineers for strategic planning while recycling for addressing environmental concerns. The book is in line with the COP26 climate summit held in Glasgow (2021).

Dr. Ranvijay Kumar
Mohali, India
Dr. Rupinder Singh
Chandigarh, India

Editors

Rupinder Singh is a professor in the Department of Mechanical Engineering, National Institute of Technical Teacher Training and Research, Chandigarh. He has received a Ph.D. in mechanical engineering from the Thapar Institute of Engineering and Technology, Patiala. His areas of research are additive manufacturing, composite filament processing, rapid tooling, metal casting, and plastic solid waste management. He has co-authored more than 350 science citation-indexed research papers, 10 books, more than 100 book chapters, and has presented more than 100 research papers in various national/international journals. His research has been cited more than 8,600 times with an H index of 44 and Google scholar i-10 index of 175. As per Standford University, he has been listed as one of the world's top 2% of scientists (data-update for "Updated science-wide author databases of standardized citation indicators", August 2021).

Ranvijay Kumar is an assistant professor in the University Center for Research and Development, Chandigarh University. He has received a Ph.D. in mechanical engineering from Punjabi University, Patiala. His areas of research include additive manufacturing, shape memory polymers, smart materials, friction-based welding techniques, advance materials processing, polymer matrix composite preparations, reinforced polymer composites for 3D printing, plastic solid waste management, thermosetting recycling, and destructive testing of materials are the skills of Dr. Ranvijay Kumar. He has co-authored more than 50 research papers in science-citation index journals, 45 book chapters, and has presented 26 research papers in various national/international-level conferences. In 2020, he was awarded the silver medal by the prestigious CII MILCA for his excellent research contribution to society.

Contributors

Rupinder Singh
National Institute of Technical Teachers Training and Research
Chandigarh, India

Ranvijay Kumar
University Center for Research and Development, Chandigarh University
Chandigarh, India

Yang Wei
Nottingham Trent University
Nottingham, United Kingdom

Inderpreet Singh Ahuja
Punjabi University
Patiala, India

Raman Kumar
Chandigarh University
Chandigarh, India

Amrinder Pal Singh
Panjab University
Chandigarh, India

Ajay Batish
Thapar University
Patiala, India

Jaspreet Singh
Lovely Professional University
Phagwara, India

Deepika Kathuria
University Center for Research and Development, Chandigarh University
Chandigarh, India

Jasgurpreet Singh
Chandigarh University
Chandigarh, India

Kamaljit Singh Boparai
Maharaja Ranjit Singh Punjab Technical University
Bathinda, India

Vinay Kumar
Guru Nanak Dev Engineering College
Ludhiana, India

Nishant Ranjan
University Center for Research and Development, Chandigarh University
Chandigarh, India

Ravinder Sharma
Thapar University
Patiala, India

Balwant Singh
Chandigarh University
Chandigarh, India

Abhishek Kumar
National Institute of Technical Teachers Training and Research
Chandigarh, India

Monica Bhattu
University Center for Research and Development, Chandigarh University
Chandigarh, India

Sanjeev Kumar
Panjab University
Chandigarh, India

Kapil Chawla
Lovely Professional University
Phagwara, India

Contributors

1 Introduction to Circular Economy and Recycling Plastics

Deepika Kathuria and Monika Bhattu

1.1 INTRODUCTION

Circular economy (CE) is a perception of integrating economic and environment systems in a sustainable way that focuses on "make/remake-use/reuse" (Sikdar, 2019). The key principles of this model are: i) product design to minimize plastic waste and pollution; ii) retain use of materials and products; and iii) natural system regeneration. The 5Rs of circular economy are reduce, repair, reuse, refurbish, and recycle. The circular economy model is a substitute for the linear economy model, primarily based on the take, make, use, and dispose (take-make-use-dispose, Figure 1.1). It means to take the raw material for making goods and, after use, dispose the waste, which will pollute the environment. This leads to numerous environmental challenges, such as environment pollution and depletion of resources. In the case of a circular economy, the plastic waste will be used to produce beneficial products or raw material for human use (Sikdar, 2019, Merli et al., 2018, Nishida, 2011, waste narratives (2016) and Ellen-Mac Arthur Foundation-vol.1).

In 2018, the concept of circular economy was promoted officially by the World Economic Forum (WEF). The WEF published the Platform for Accelerating the Circular Economy (PACE). in association with many organizations such as Ellen MacArthur Foundation, World Resources Institute (WRI)., to endorse the circular economy. Recently, this idea was included in the European Union Horizon 2020 strategy. As per the Ellen MacArthur Foundation, "A circular economy is one that is restorative and regenerative by design and aims to keep products, components, and materials at their highest utility and value at all times, distinguishing between technical and biological cycles" (Ellen-Mac Arthur Foundation-vol.1).

Basically, circular economy works at four levels, i.e. products, companies, networks, and policies. These levels can be achieved by various means: a) recyclable and reusable product should be designed, which is based on manufacturing using clean methods and on green supply chains (WRI initiatives). In 2016, industrial parks in China prevent the release of greenhouse gases of about 14 million tonnes by plastic recycling (Liu et al., 2018).; b) companies/ industries should establish innovative business models that will help in creating

DOI: 10.1201/9781003184164-1

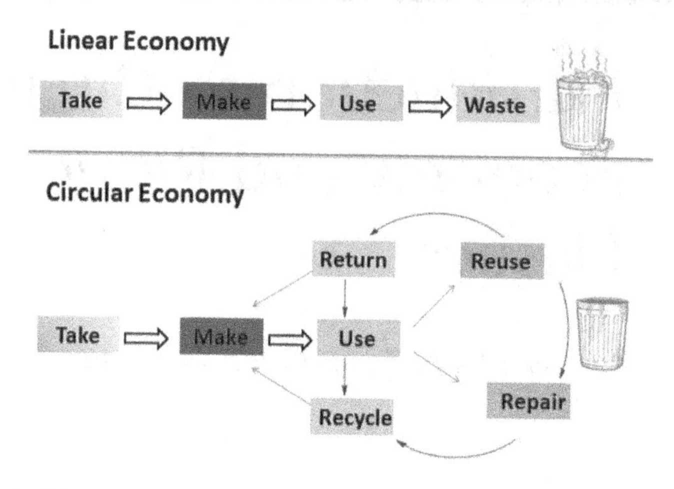

FIGURE 1.1 Diagrammatic representation of linear economy and circular economy.

and maintaining the private and public values; c) network of industries that produce key materials and customers that consume those products should be interconnected; d) implementation of policies that supports the market directly or indirectly (Geng et al., 2019). The above levels must be included in a worldwide circular economy strategy that will mainly help to reduce inputs of material, labor, energy, and emission of carbon footprints (Ellen-Mac Arthur Foundation-vol.1). A report by the Ellen MacArthur Foundation suggests that if all the waste material is used in reverse-cycle activities, the net cost savings of material used can be USD$340 to $380 billion p.a. and USD$520 to $630 billion p.a. at EU level for "transition scenario" and "advanced scenario," respectively. The circularity is greatly influenced by many factors: i) the type of material used to make plastics; ii) the type of plastics used; and iii) the sociopolitical framework in which plastic waste is generated, consumed, and managed. Taking all these things into consideration, it has become an issue of utmost importance to have the knowledge of the different methods of the recycling of plastic waste and different kinds of circular economy approach that are already implemented that will help the future researchers to work towards a more efficient recycling processes and towards a more sustainable economy.

1.2 PLASTIC AND ITS WASTE

Plastics are one of most useful industrial inventions of the modern world, and have been utilized in various sectors consisting of electrical and electronic equipment (EEE), construction, automotive, packaging, health care, and agriculture. Due to these applications and the lightweight and durability of many plastic-based products have been manufactured in the 20th century whose negative consequences we are now facing in 21st century as it is polluting our nature (Thompson 2009). The production of plastic/polymer has been drastically increased from 2 million tonnes in 1950 to 381 million tonnes in 2015. In total,

7.8 billion tons of plastic had been produced by the world by 2015. Unfortunately, only 20% of the plastic was recycled and the rest of the plastic was incinerated (25%) or discarded (55%) (Our world indata). Moreover, in 2015, the CO_2 emission due to the use of conventional plastic was 1.7 Gt CO_2-eq and under the current trajectory it will increase to 6.5 Gt CO_2-eq by 2050 (Zheng & Suh, 2019). Some projections reported that the amount of plastic waste accumulated in oceans may increase to triple that by 2025, if no initiative is taken, and if the linear economy continues by 2050 we will end up with more plastic by weight than fish in the oceans. As a result, plastic waste is now recognized as a serious solid waste management problem (Rebeiz & Craft, 1995). This possesses serious consequences to human health and environment. The degradation of plastic is a time-consuming process and, as a result, it remains in the environment for a long time, causing damage to the ecosystem and biodiversity and thus causing social and economic problems (Boucher & Friot, 2017). Plastics are also responsible for physical and toxicological impacts due to their ingestion or inhalation by biota and this leads to damage in the food chain (Thompson, 2015), due to bioaccumulation and chronic exposure (Wright & Kelly, 2017).; thus, management of plastics is crucial. Various factors such as infrastructure, poor waste management infrastructure, lack of public awareness, and application of insufficient recycling technologies play a major role in accumulating the plastic waste in the environment and making it pervasive (Jambeck *et al.*, 2005 and Jambeck et al., 2017); Hence, efficient management of plastic waste has become an important criterion to achieve circular economy.

Generally, five types of plastic contribute to producing plastic wastes, which include 1) polyethylene terephthalate (PET) – used for beverage and food packaging; (2) low-density polyethylene (LDPE) – used in plastic bag production, tubing, etc.; (3) high-density polyethylene (HDPE) – used for milk, shampoo, and detergent bottles; (4) polypropylene (PP) – used mainly in the automotive industry; and (5) polystyrene (PS) – used for food packaging. The polymers are tabulated in Table 1.1 with their application and recyclability potential.

In September 2015, the Sustainable Development Goals (SDGs). of the United Nations (UN) was agreed to by 193 UN member-states. Various targets were included that were relevant to the waste and resource management with the goal to improve sustainable waste and resource management directly or indirectly (Pedersen, 2018).

In 2017, the assembly third session (UNEA-3) emphasized the importance of establishing collaborative relationships between governments, regional bodies, the private sector, particularly major commercial actors, civil society, nongovernmental organizations, all relevant international and regional organizations, and conventions for adopting a better waste management system to address the issue of marine litter (Finska, 2018). As a result, increased emphasis is needed not only on reducing overreliance on plastics but also on promoting their circularity potential in the system as this is the basic way of decreasing the environmental, economic, technical, and social implications of plastic waste. Further, the most efficient way to minimize the disposal of plastic waste material is *recycling,* which will boost the circular economy.

TABLE 1.1
The list of polymers with their applications and recycling status

Polymer	Applications	Current recycling status
Polyethylene terephthalate (PET/PETE)	Soft drink and water bottles, fruit juice container, cooking oil, oven-ready meal trays, and food jars.	Very commonly recycled
High-density polyethylene (HDPE)	Containers (personal and home care products), carbouys and cans for edible oil, milk and lube oil and shampoo bottles, and also bottles for personal care products.	Very commonly recycled
Polyvinyl chloride (PVC)	Blood and urine bags in medical field, doors, windows, packaging (shrink, wrapping, labeling, blister packaging, and cling packaging)., rain wears, flexible holdings, and pipes.	Commonly recycled
Low-density polyethylene (LDPE)	Bubble foil, wrapping purpose, grocery bags, and squeezable bottles.	Sometimes recycled
Polypropylene (PP)	Medicinal bottles, chip packs, bottle caps, ketchup bottles, furniture, butter tubs, and consumer luggage.	Commonly recycled
Polystyrene (PS)	Compact disc jackets, cutlery, disposable cups, and packaging foam.	Typically not recycled
Miscellaneous category that includes polycarbonate, thermoset plastics, PUF Bakelite, multilayer and laminated plastics, melamine, nylon, fiberglass, acrylonitrile butadiene styrene (ABS)	Clothing, defense gadgets, electronic goods, food packaging, nets, and ropes.	Typically not recycled

1.3 PATHWAYS FROM WASTE TO MATERIALS: WAYS OF RECYCLING

The circular economy has brought a revolutionary shift in the management of plastic waste management. The recycle of plastic is the key way to achieve circular economy (Schyns and Shaver, 2020). Recycling of plastic waste generally has four types: primary recycling, secondary recycling, tertiary recycling, and quaternary recycling and these plastic recycling fates are as shown in Figure 1.2.

1. *Primary Recycling*: Primary recycling is very common way of reprocessing that involves the cleaning or semi-cleaning of recovered products and their reuse for its original application. An example of primary recycling is the use of reused PET bottles from postconsumer and to use them

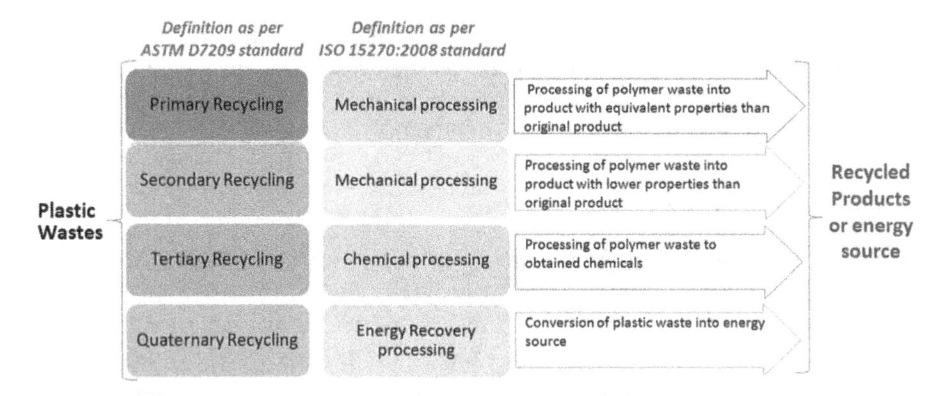

FIGURE 1.2 Different types and processing of plastic recycling.

for the production of new bottles. The main drawback of this process is that transport of disposed plastic is neither economically viable nor technically feasible as various processes like segregation of source, collection, and sorting of disposed waste involved in the postconsumer waste management process chain before its transportation into production (Hahladakis et al., 2018).

2. *Secondary Recycling:* Secondary recycling is the most widely used method for recycling and is considered economically viable for foams and rigid plastics. It is also known as *mechanical recycling* (Meran et al., 2008 and Brachet et al., 2008). This process consists of pretreatment, decontamination, and reprocessing of sorted plastic waste involving the following steps: (a) cutting and shredding of plastic waste into small flakes; (b) decontamination by other materials (e.g. paper, dust, etc.) using a cyclone; (c) density-based separation by floating; (d) milling; (e) washing with water and chemicals for the removal of adhesives; (f) drying; (g) agglutination using pigments or additives; (h) cast into strands and pelletization; and (h) quenching using cold water in order to make it granulated and solid. Preparation of flooring tiles from mixed polyolefins is an example of secondary recycling.
3. *Tertiary Recycling:* Tertiary recycling is also known as chemical recycling. It is a process in which the collected plastic waste is used as a feedstock for basic chemicals and fuels (Garcia, 2016 and Tukker, 2002). This process involves various methods like solvolysis, thermolysis (Oliveux et al., 2015), microwave irradiation (Achilias et al., 2010), pyrolysis, high temperature effect, gasification, thermal cracking, glycolysis, photodegradation, hydrolysis, methanolysis, and chemolysis (Singh et al., 2017). The formation of diols and dimethyl terephthalate by glycolysis of PET is an example of tertiary recycling.
4. *Quaternary Recycling*: Quaternary recycling, also known as energy recovery, is a technique that involves incineration to recover the energy component of plastic waste. Combustion with energy recovery is the most

effective technique of reducing waste volume, especially for low-quality, mixed, and severely contaminated plastic waste. As open burning has various negative health and environmental impacts so it is not considered in context with the CE (Song et al., 2009).

1.3.1 Mechanical Recycling

The mechanical recycling process involves the recovery of plastic from plastic waste resource which can be used to generate similar or different products. European standard such as EN15342 to EN 15348 are available for recycling plastics such as PE, PVC, PET, and PS (EN 15342 to EN 15348). For the recycling of conventional plastics such as like PET, PE, PP, and polystyrene mechanical recycling are the preferred option (Hopewell et al., 2009). Prior to recycling, plastics wastes are sorted according to their resin type and their chemical characteristics. Various sorting systems are reported to determine the resin, which varies from manual sorting and picking of plastic materials from waste to mechanized automation processes that involves sieving, shredding, or separation by density. The magnetic spectrophotometric distribution technologies like UV-visible spectroscopy, near-infrared spectroscopy, laser, fluorescence spectrophotometry, XRD, etc. are also used for this process (circulareconomyasia.org). Some plastic materials are also separated on the basis of color before they are recycled, which will be furthered proceeded for washing for the removal of organic contaminants. After sorting, the next step involves grinding plastic products into flakes followed by compounding and pelletizing that coverts flakes into granules, which are easy to process. As per the CE, the mechanical recycling of material is distinguished into open-loop recycling and closed-loop recycling.

Closed-loop recycling, often known as "upcycling or horizontal recycling," keeps the recovered plastic materials designed and manufactured qualities the same as their virgin counterparts. As a result, the recycled plastic can be utilized to make the same products as before. However, with the use of different ingredients, the product produced can be completely new. Recycling of PET bottles into new bottles is as an example of closed-loop recycling (Ahvenainen, 2003). Polymers such as polypropylene (PP) and high-density polyethylene (HDPE) can be recycled through this process as the properties remain intact in the manufacturer's products as of the original product (www.exeley.com). On the other hand, the designed and manufactured features of recycled plastic materials are degraded to use the material to generate other products which may be of lower-quality goods than the ones from which they were originally recovered. This process is known as open-loop recycling, also termed "downcycling" or "cascading" (Eriksen et al., 2019). Formation of PET fibers from PET bottles illustrates an example of the open-loop recycling process. Figure 1.3 depicts the closed- and open-loop recycling of PET bottles. Unfortunately, the currently known mechanical recycling processes are limited due to the following reasons: a) cost, b) inconsistency in the product quality, and c) mechanical properties degradation.

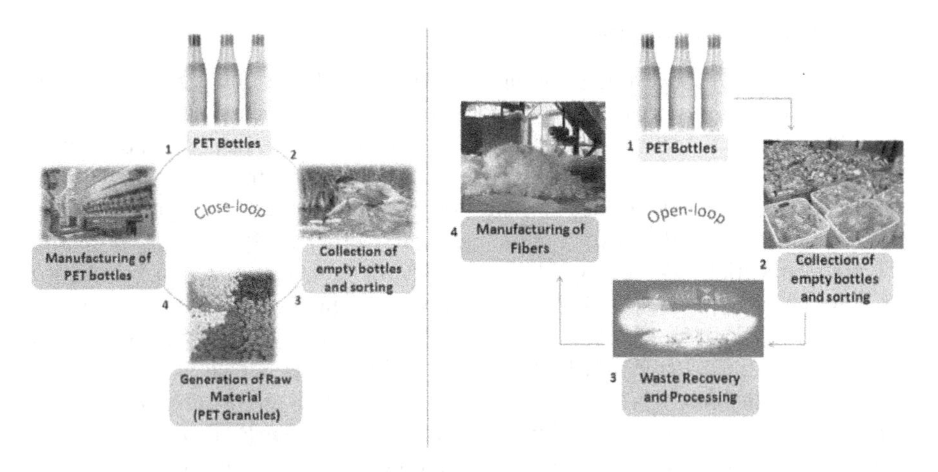

FIGURE 1.3 Recycling of PET bottles via closed-loop and open-loop approach.

1.3.1.1 Industrial Approaches for Mechanical Recycling of Plastic

The most common example of plastic recycling in industries is the recycling of PET. This process consists of collection, sorting, and reprocessing. The European Union Directive of Packaging and Packaging Waste (2004/12/EC) is responsible for handling the PET collection in Europe (Directive 2004/12/EC of the European Parliament and PPW_report_2010–2012). The collection method includes curbiside collection, location refill, and deposit system. The collection process followed depends upon the amount of contamination present and collection quantities. After collection, sorting of PET is done from contaminating plastic waste like HDPE and PVC either by automated systems or by the hand-sorting method. This sorted plastic is further compressed into bales for transportation and finally processed into new products like bottles, containers, or into fibres or for any new applications (Awaja & Pavel, 2005 and Ragaert et al., 2017). The most employed technique for the mechanical degradation of PET is solid-state post-condensation. In this process, PET is heated between the melting temperature and glass transition temperature. The condensation reaction takes place in an amorphous phase of polymers between two terminal groups at a temperature ranging from 200–240°C (Welle, 2011). The chain extender utilized in this process is Joncryl, a multifunctional epoxy-based oligomer from BASF (Baden Aniline and Soda Factory) (Awaja & Pavel, 2005).

In the case of waste electrical and electronic equipment (WEEE), acrylonitrile butadiene styrene (ABS) is the most commonly used recycled polymer. The mechanism of recycling involves both cross linking and chain scission. But the recycling of ABS leads to weakening of mechanical properties out of which impact strength and ductility are highly affected. In order to improve these properties, an upgrade using virgin ABS blending and the addition of impact modifiers or chain extenders is done (Arostegui et al., 2006, Peydro et al., 2013, Scaffaro et al., 2012 and Bai et al., 2007).

1.3.2 Chemical Recycling

The chemical recycling strategy is an approach to achieve circular economy by reducing the raw material demand and negative environmental impacts (Salem *et al.*, 2009). This recycling strategy involves either depolymerization of polymer or repurposing of waste. In the depolymerization technique, the polymer waste is depolymerize returning to their starting point under specified settings feedstocks that have been purified and then repolymerized in order to produce virgin-quality polymeric materials (Oliveux et al., 2015). The depolymerization can be chemolysis, thermal (pyrolysis), hydrolysis, glycolysis, or photodegradation. The schematic representation of depolymerisation of PET by chemolysis using methanol and water is given in Scheme 1.1. Methanolysis of PET gives dimethyl terephthalate (DMT) and ethylene glycol, whereas upon hydrolysis it gives terephthalic acid (TPA) (Han, 2019). As discussed previously, post-consumer PET can be recycled mechanically, however it often leads to the production of lower-quality product, whereas chemical recycling retains the quality of the product.

Polycarbonate (poly(bisphenol A carbonate, PC) is a thermoplastic polymer that has wide applications in electronics, automobiles, containers, optics, etc. due to their high-impact resistance, flame retardance, low cost in production, and good optical clarity. Only a small amount of PC has been recycled mechanically, which eventually goes to the landfill or incineration because of eventual loss of its properties. The chemical recycling of PC is gaining much attention from researchers. The depolymerisation of PC can be performed by pyrolysis, hydrolysis, alcoholysis, aminolysis, or reduction (Scheme 1.2). The incomplete recycling of feedstock during the depolymerization process is still its major drawback (Kim, 2020, Geyer et al., 2017 and Jones et al., 2016).

Another approach for chemical recycling is the repurposing method, where polymer waste is converted into building blocks for the generation of new polymeric materials of high value (Figure 1.4). An example of repurposing of plastic waste is the conversion of polycarbonate (PC) into PSU (poly(aryl ether sulfones),

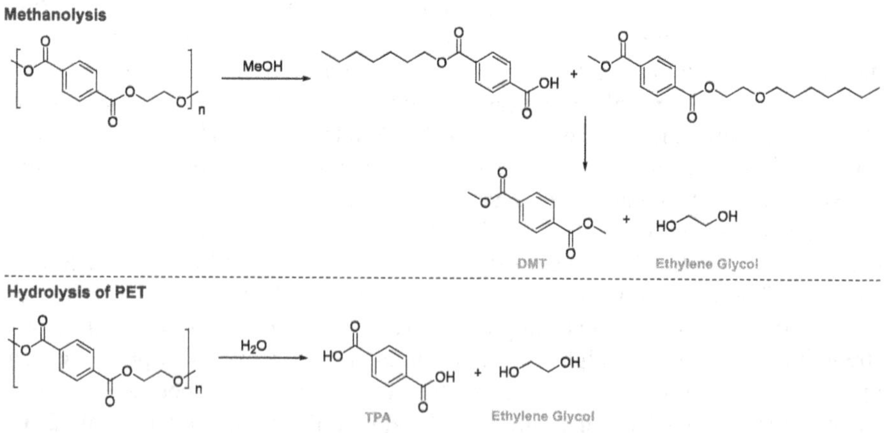

Methanolysis

MeOH

DMT

Ethylene Glycol

Hydrolysis of PET

H_2O

TPA

Ethylene Glycol

SCHEME 1.1 Depolymerisation of PET via methanolysis and hydrolysis.

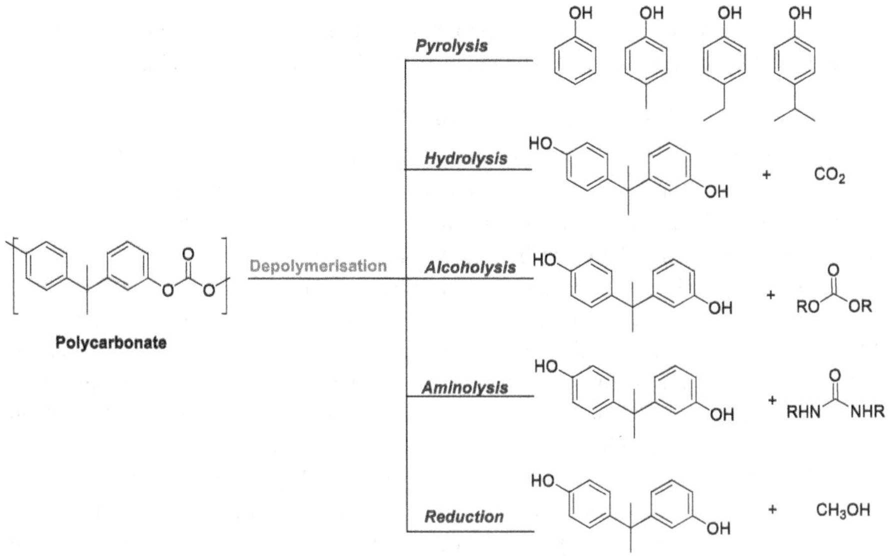

SCHEME 1.2 Ways to depolymerise polycarbonate.

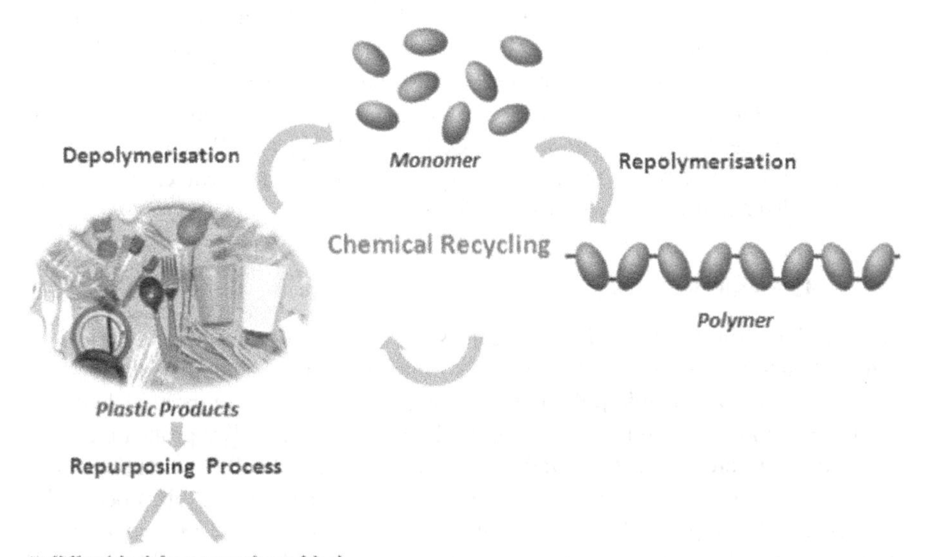

FIGURE 1.4 Approaches toward chemical recycling.

a highly used engineered thermoplastic for reverse osmosis, water purification, and high-temperature applications (Jones et al., 2016). As PC materials are not recycled completely by conventional mechanical recycling, as a result, a ton of PC waste is often dumped and ends up in landfills after use. So, chemical recycling has proved much more effective in terms of the circular economy. Another example of the repurposing process involves the polymer degradation for its conversion to high-value products such as liquid fuels and waxes (Hong & Chen, 2017).

1.3.2.1 Industrial Approaches for Chemical Recycling of Plastic

Various industries are using the chemical recycling methods for recycling plastic wastes, which will help in boosting the circular economy (ClosedLoopPartners, 2019). A method for the chemical recycling of polypropylene (PP) waste was developed and patented, where the first step is purification to remove color and contaminants and which will undergo chemical disintegration in a solvent at 80–220°C temperature and 150–15,000 psig of pressure to obtain the pure PP (Layman et al., 2017). The process of recycling of polyethylene (PE) i.e. HDPE and LDPE is patented by BioCellection, Inc. where they have utilized an accelerated thermal oxidative decomposition (ATOD) process. This process requires treatment of PE with an oxidation agent such as HNO_3 at 60–200°C for a period of 30 min–30 h, which will break the polymer into various organic compounds such as adipic, azelaic, glutaric, pimelic, succinic, and suberic acid, which can be purified further (Yao et al., 2019).

Agilyx, polystyert, and pyrowave have developed a method for recycling polystyrene (PS) into its monomer i.e. styrene using pyrolysis. Agilyx uses the high temperature (350–500°C), whereas the pyrowave process uses a sequential working temperature of 55–90°C followed by 2–10°C. Interestingly, the yield of monomer can be obtained up to 95%, which can be repolymerised to get virgin PS. In the case of the polystyert process, p-cumene was used to purify the PS from PS wastes and the used solvent was recovered later on. Various companies such as Plastic2Oil, Resynergi, GreenMantra Technologies, Cadel Deinking, and Plastic Energy are focusing on the recycling of mixed plastic wastes to obtain fuels (Bauer et al., 2020, Gil et al., 2017 and Fullana & Lozano, 2013).

1.3.3 Organic Recycling

The circular economy basically emphasizes the environmental and economic sustainability and one of the best measures for obtaining the sustainability of polymers is the use of biodegradable polymers such as PLA (poly lactide), poly (3-hydroxy butyrate), cellulose-based plastics, lignin-based plastics, polycaprolactone, etc. Such polymers are suitable for various applications, specifically for one-time use i.e. packaging materials and various other utilities in agriculture. In this process, the plastics are decomposed by microorganisms into CO_2, CH_4, H_2O, and biomass (microbial population growth). Compositing or advanced biological treatment (i.e. AD processes) is involved in the process of organic recycling. However, to implement this process, a separate collection of biodegradable plastic waste is required (Payne et al., 2019). Keeping this thing in mind, European standard EN 13432 reported some strict strategies regarding the organic recycling of compostable, biodegradable waste. Plastic Europe, an association of plastic manufactures, reported the organic and mechanical recycling methods of plastic waste as the most effective methods for plastic waste remediation (www.plasticseurope.org). Although both chemical and biological recycling is considered "green," thorough and objective life cycle analyses are required to assess their long-term viability.

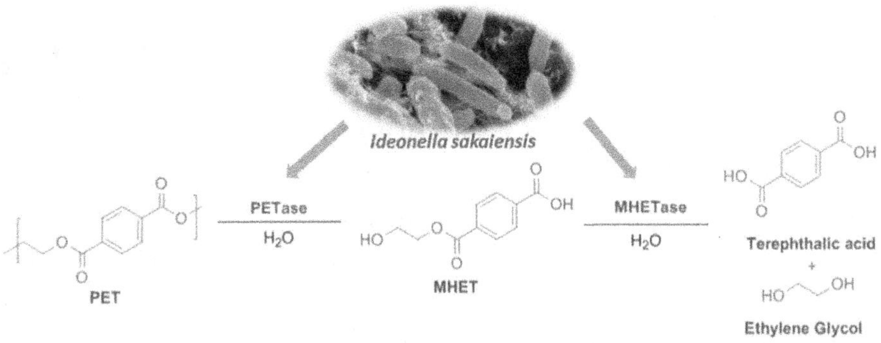

SCHEME 1.3 Munching of PET using *Ideonella sakaiensis.*

Interestingly, scientists are working towards the development of methods for organic recycling of conventionally considered non-biodegradable plastics such as polyethylene and PET wastes as well. For example, Yoshida *et al.* identified the *Ideonella sakaiensis* strain 201-F6 (a bacteria), which can utilize poly (ethylene terephthalate) (PET) for carbon and energy sources (Yoshida et al., 2016). *Ideonella sakaiensis* contains a plastic munching enzyme i.e. PETase and MHETase, which breaks PET into its building blocks such as terephthalic acid and ethylene glycol (Scheme 1.3), which can be utilized as building blocks for the preparation of PET. Many oxidoreductase have been shown to possess the capability to degrade PE. The development of engineered enzymes for plastic degradation can open up a new era for organic recycling of plastic wastes (Wei & Zimmermann, 2017). Recently, an engineered PET depolymerase enzyme has been reported with an improved ability of PET hydrolysis into its monomers (Austin et al., 2018 and Tournier et al., 2020).

1.3.4 Energy Recovery/Incineration

Among all the plastic waste recycling options, this method is the least preferred choice in the CE as it completely destroys the plastic material. Energy recovery from plastic waste is the conversion of non-recyclable plastics into useful energy i.e. electricity, heat, or fuel, which can utilized for various technology processes (Figure 1.5) (Ignatyev et al., 2014). For the recovery of energy content, their burning is preferred specifically for the heavily contaminated and mixed plastic waste that are very difficult to recycle or the recycling is not environmentally and economically feasible. The process of energy recovery includes various processes like gasification, combustion, anaerobic digestion, pyrolization, and landfill gas recovery. The energy recovery processes are generally distinguished between two types of techniques: mass burn and refused derived fuel systems. In mass burn, the waste material is combusted without any pretreatment and in the other method the material is used after the sorting process and possesses defined quality parameters. After that, the energy is produced from waste by giving it thermal treatment. Finally, the released energy is converted into the transportable form of energy

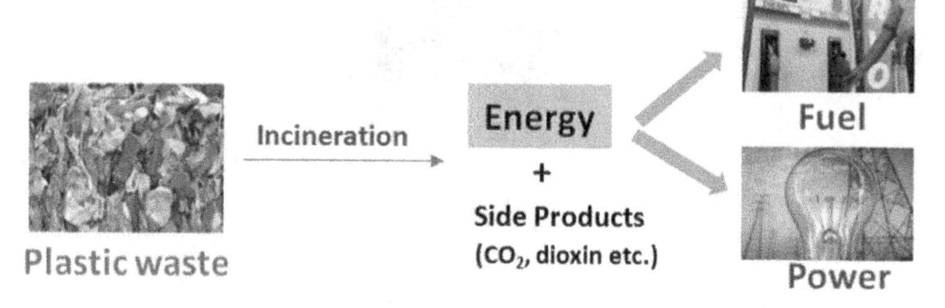

FIGURE 1.5 Energy recovery process using plastic waste.

i.e. heat, fuel, and electricity and the clearing of emissions has been done to ensure the safety of the environment from waste gases (docs.european-bioplastics.org). By incarnation of plastic with the saving of energy ~ 30% of the carbon footprint gets reduced when compared to the use of other sources of carbon. The energy recovery system utilized most prominently around the world is the municipal solid waste (MSW) combustor or waste-to-energy (WTE) plant. Such plants combust MSW on a massive scale to produce steam, hot water, or electricity depending on the market demand (Fisher et al., 2005).

Using this method, the volume of garbage is reduced by 90–95% but incineration also imposes several challenges due to release of substances like volatile organic compounds, polychlorinated dibenzofurans, dioxions, and polychlorinated biphenyls, which have deteriorating impacts on the environment and human health (Singh et al., 2017 and Valavanidis et al., 2008). However, many technologies have been developed that provide ways for non-toxic emission. From the incineration of plastic waste, two types of ashes i.e. fly ash and bottom ash are produced. Fly ash is composed of fine particles present in the incinerator exhaust gas, whereas the bottom ash is present at the bed of the incinerator as it is composed of heavy and large particles. Their disposal is another issue in the process of incineration as landfilling of these ashes can contaminate the groundwater and soil (Siddique et al., 2008).

1.4 IMPLEMENTATION OF CIRCULAR ECONOMY FOR PLASTIC WASTE: CHALLENGES AND OPPORTUNITIES

The circular economy for plastic has become resource efficient and has proven very helpful in diminishing the use of plastic waste in the generation of plastic from renewable resources. As defined previously, it is evident that the designed concept of circular economy was developed from the sustainable economy approaches. According to current estimates, the contribution of the circular economy towards the global economy is USD$1 trillion/year, which leads to the assumption that economic benefits can be potentially huge if a circular economy is completely implemented. To meet the needs of a circular economy approach, various changes in the current practices are required to be employed, which involves sustainable approaches to reuse, maintenance, eco-design, recycling, leasing, sharing, and chemical conversion

(Kovacic et al., 2019 and Iaquaniello et al., 2018). As discussed previously, various initiatives like the use of bioplastics and recycling of plastic waste, etc. have been taken by different countries for plastic waste that encourages the sustainable management of plastic waste and helps the world to take a move towards a circular economy.

In order to manage the e-waste, the extended producer responsibility (EPR) scheme (Lifset, 1993 and Lindhqvist & Lifset, 2003) was introduced according to which grouping and recycling of plastic waste is migrated from government organizations to producers (www.corpseed.com, Kunz et al., 2018 and Filho *et al.,* 2019). The basic objective of this scheme is to take favorable steps towards the circular economy by incentivizing the design of a product to diminish the waste and harmful chemicals. However, it is difficult to implement EPR at the level of producers to achieve a circular economy at the worldwide level. In order to solve this issue, stakeholders are required in EPR to work with producers, consumers, municipalities, recyclers, etc (Kunz et al., 2018). The implementation of EPR in Europe supported the transition towards a circular economy, which can be clearly seen with the example of a reduction of 35% e-waste volume in 2012. In December 2015, the European Commission launched an ambitious circular economy action plan to combat climate change and the environment while also increasing economic growth, investment, social equity, and job creation. The EU continues to encourage activities that lead to a circular economy, such as the EU Plastics in a Circular Economy Strategy (European Commission, Plastic Waste). But to achieve a circular economy in plastics, massive investments in people and infrastructures are required; however, the potential economic rewards are substantial. In spite of these initiatives and schemes, it is possible that there is not enough collective approaches and international cooperation that makes a significant difference. Various challenges have been faced in implementing a circular economy. However, these challenges when observed may also be helpful in creating vast social, economic, and environmental opportunities in the fields of plastics and other waste management industries (Bucknall, 2020).

The existing challenges to begin a plastic circular economy involve the economic, cultural, technical, and social problems. Technical issues involve the collection and sorting of plastic waste that should be taken care of to enhance the process of recycling. The separation of additives from the waste material also becomes a burning challenge in a circular economy approach. For this purpose, various chemical processes were proposed, but these processes have some limitations, such as cost, trained personnel, and high-cost equipment, are also needed. Another important issue is that over time the chain length degradation of plastic occurs, which leads to an impossible use of that plastic for its original application. Upon repeated cycles of recycling, the loss of inherent properties of plastic occur.

Apart from it the major challenge in achieving the CE is the implementation of policies and regulations and how these can be approached to the consumers and producers. Hence, it has become more important to make the required changes in the policies such that everyone is aware of them. Nowadays, mechanical recycling is very unorganized and informal, which leads to the emission and leakage of hazardous chemicals in environment. To implement a CE, the labeling should be

TABLE 1.2
The representative symbols for recycling with their specifications

Symbol	Specification
	This indicates that 75% or more local governments collect packaging for recycling.
	Packaging collected by 20–75% of UK local authorities for recycling
	Packaging collected by less than 20% of local authorities for recycling
	Symbol used across the EU that indicates that the producer has made a financial contribution to packaging recovery and recycling in Europe
	Each mobius loop is accompanied by a number (1–7) and letters for identifying the polymer type.
	Product is industrially biodegradable to a European standard
	Product is appropriate for home composting

proper in order to avoid confusion among individuals. Due to incomplete knowledge or improper label, people are unsure of how to cope with plastic waste in a proper manner i.e. whether to put in a waste bin, recycle bin, or in a compost heap. Labels should be simple, clear, and self-explanatory. Some representative symbols for recycling are as shown in Table 1.2, with the specifications (www.ellenmacarthurfoundation.org).

1.5 PRACTICES TOWARDS THE CIRCULAR ECONOMY APPROACH

The first major approach to achieve CE is the reuse of the plastic, which will help to reduce the plastic pollution. For enhancing the thinking of reuse of plastic, New Plastics Economy started working on reuse models in 2019, which shows various advantages of the reuse of both to the consumers and businesses (depositreturnscheme.zerowastescotland.org.uk). For adopting the techniques for reuse, the strategies should be made by policy makers and manufacturers which will help to raising awareness and directing consumers.

Another approach to achieve CE at the consumer level is the execution of the deposit-refund system as it enhances the efficiency of collection of plastic waste.

The deposit-return scheme is implemented in Scotland to enhance the recycling of used bottles (especially PET bottles). This scheme is designed based on the deposit of 20p on buying a single-use container drink and when the consumer returns the empty bottle he/she will get back the deposit money (futurenviro.es/en/el-sistema-de-deposito-devolucion-y-retorno-aleman-sddr). But this approach in Scotland is only for PET, glass, and aluminium bottles. Germany and 40 other regions had already introduced this system where on deposition of the container, the monetary value will be provided to the consumer. For a returnable container, which can be used many times after washing, the associated value is between €0.08 and €0.15, whereas for a recyclable container (single-use, can be recycled for other purpose) the value is €0.25. Such strategies should be implemented to boost the collection and sorting process and thus adopting such a scheme across the world will help to improve the circularity of plastics (depositreturnscheme.zerowastescotland.org.uk and futurenviro.es/en/el-sistema-de-deposito-devolucion-y-retorno-aleman-sddr).

The next approach should be practice in designing products that can be easily recycled. The use of mixed polymers to make the product more attractive to consumers and to provide a unique property to the product creates a problem in recycling. So, the development of an eco-design product that meets the social, environmental, and economic aspects without affecting the standards of the product is important. To meet the challenges of product design, the following principles need to be implemented: a) eco-design, b) design for sustainability, c) remanufacture design, and d) life cycle thinking (Simon, 2019, Spangenberg et al., 2010 and Charter, 2018).

1.6 FUTURE PROSPECTIVE

Undoubtedly, plastics are not the only materials that have versatile applications but also, they are also an important resource and it appears that they are becoming more precious. Given the massive impact of plastic pollution on life on Earth, as well as the contributing factors, we require a new economic paradigm to govern investment, production, and consumption decisions. According to the World Economic Forum and the Ellen MacArthur Foundation (2017), indicates a New Plastic Economy based on the premise that plastics never become garbage, rather they can be used again and again. As a consequence, the circular economy's objectives and ambition to yield superior economic and environmental outcomes are matched with the new plastics economy, which aims to substantially reduce the plastics loss in natural systems (especially the ocean) and decouple from fossil raw materials (Paletta et al., 2019). The following areas emerge as potential research priorities for technical innovation as a result of the preceding discussion:

- Growth of cost effective and environment friendly "chemo- and biocatalytic recycling techniques." As some techniques are already well progressed like "biocatalytic recycling of polyesters."
- Design for circularity: new plastics must be easy to disassemble and recycle in a closed loop.

- Create bio-based plastics that are recyclable and biodegradable once they've served their purpose. In this regard, PHAs as prospective polyolefin substitutes offer a lot of promise (Sheldon & Norton, 2020).
- It may be possible to build efficient microbial communities capable of degrading plastic trash, including varieties that are now resistant to biological breakdown. The combination of mechanical, chemical, thermochemical, and biotechnological recycling processes with microbial, fungal, or any other biological activity allowed to occur under regulated and contained conditions could be the key to achieving the circular economy goal in this area.

Besides these, it may also be claimed that beyond a better knowledge and optimization, there are few technological obstacles left to address, at least at the laboratory level, and that many of the larger obstacles are driven by economic forces and social behavior. To put it another way, the issue is not with plastics; it is with humans and what they do with them. Hence, a change in behavior is also a key component of the solution.

1.7 SUMMARY

To summarize, the plastic waste has become the most burning topic of today's society. Various implementations have already been done to the recycling of plastic waste. The circular economy has played an imperative role in the recycling of plastic waste and towards the sustainable economy growth. In this chapter, the author compiled the basic objectives of the circular economy towards the recycling of plastic, initiatives taken towards the execution of circular economy, and the various ways of recycling of the plastic waste that comes with a vital role in sustainable economic growth. The author also highlighted some future perspectives of the circular economy.

REFERENCES

Achilias, D. S., Redhwi, H. H., Siddiqui, M. N., Nikolaidis, A. K., Bikiaris, D. N., & Karayannidis, G. P. (2010). Glycolytic depolymerization of PET waste in a microwave reactor. *Journal of Applied Polymer Science, 118*, 3066–3073.

Ahvenainen, R. (Ed.) (2003). *Novel food packaging techniques*. Elsevier.

Al-Salem, S. M., Lettieri, P., & Baeyens, J. (2009). Recycling and recovery routes of plastic solid waste (PSW).: A review. *Waste Management, 29*, 2625–2643.

Al-Salem, S. M., Lettieri, P., & Baeyens, J. (2010). The valorization of plastic solid waste (PSW). by primary to quaternary routes: From re-use to energy and chemicals. *Progress in Energy and Combustion Science, 36*, 103–129.

Arostegui, A., Sarrionandia, M., Aurrekoetxea, J., & Urrutibeascoa, I. (2006). Effect of dissolution-based recycling on the degradation and the mechanical properties of acrylonitrile–butadiene–styrene copolymer. *Polymer Degradation and Stability, 91*, 2768–2774

Austin, H. P., Allen, M. D., Donohoe, B. S., Rorrer, N. A., Kearns, F. L., Silveira, R. L., et al (2018). Characterization and engineering of a plastic-degrading aromatic polyesterase. *Proceedings of the National Academy of Sciences of the United States of America,115*(2018)., E4350–E4357.

Awaja, F. and Pavel, D. (2005). Recycling of PET. *European Polymer Journal, 41*, 1453–1477.
Bai, X., Isaac, D. H., & Smith, K. (2007). Reprocessing acrylonitrile–butadiene–styrene plastics: structure–property relationships. *Polymer Engineering & Science, 47*, 120–130.
Bauer B., Tanne J., Spott T., Mayhew F., McInnis A. J., Leis M., Cardinal C., Greer T. (2020). Microwave methods for converting hydrocarbon-based waste materials into oil and gas fuels (WO2020006512A1)., Resynergi, Inc., USA.
Brachet, P., Høydal, L. T., Hinrichsen, E. L., & Melum, F. (2008). Modification of mechanical properties of recycled polypropylene from post-consumer containers. *Waste Management, 28*, 2456–2464.
Boucher, J. & Friot, D. (2017). Primary microplastics in the oceans: a global evaluation of sources Gland, Switzerland: Iucn., 227–229.
Bucknall, D. G. (2020). Plastics as a materials system in a circular economy. *Philosophical Transactions of the Royal Society A, 378*, 20190268.
Charter, M. (Ed.) (2018). *Designing for the circular economy*. Routledge, New York, USA.
ClosedLoopPartners (2019). Impact Report.
Directive 2004/12/EC of the European Parliament and of the Council of 11 February 2004 amending Directive 94/62/EC on packaging and packaging waste (OJ L 47/26 of 18.2.2004).
Eriksen, M. K., Christiansen, J. D., Daugaard, A. E., & Astrup, T. F. (2019). Closing the loop for PET, PE and PP waste from households: Influence of material properties and product design for plastic recycling. *Waste Management, 96*, 75–85.
European Commission, Plastic Waste: a European strategy to protect the planet, defend our citizens and empower our industries (16 Jan 2018, IP/18/5, https://ec.europa.eu/commission/presscorner/detail/en/IP_18_5). [press release].
European Standard EN 15342 to EN 15348 on "Plastics. Recycled Plastics"l
Finska, L. (2018). Did the latest Resolution on Marine Plastic Litter and Microplastics take us any closer to pollution-free oceans? The JCLOS Blog Available at: http://site.uit.no/jclos/files/2018/01/JCLOS-Blog-100118_Marine-Litter_Finska. pdf.
Fisher, M. M., Mark, F. E., Kingsbury, T., Vehlow, J., & Yamawaki, T. (2005). Energy recovery in the sustainable recycling of plastics from end-of-life electrical and electronic products. In *Proceedings of the2005 IEEE International Symposium on Electronics and the Environment*, (pp. 83–92). IEEE.
Fullana Font A., Lozano Morcillo A. (2013). Method for removing ink printed on plastic films (WO2013144400A1)., Universidad de Alicante, Spain.
Garcia, J. M. (2016). Catalyst: design challenges for the future of plastics recycling. *Chem, 1*, 813–815.
Geng, Y., Sarkis, J. and Bleischwitz, R., (2019). *How to globalize the circular economy*. Nature Publishing Group: London, UK.
Geyer, R., Jambeck, J. R., & Law, K. L. (2017). Production, use, and fate of all plastics ever made. *Science Advances, 3*, e1700782.
Gil A., Dimondo D., Rybicki R. (2017). Reactor For Continuously Treating Polymeric Material (US20170327663A1)., Greenmantra Recycling Technologies Ltd., Can.
Hahladakis, J. N., Velis, C. A., Weber, R., Iacovidou, E. and Purnell, P. (2018). An overview of chemical additives present in plastics: Migration, release, fate and environmental impact during their use, disposal and recycling. *Journal of Hazardous Materials, 344*, 179–199.
Han, M. (2019). Depolymerization of PET bottle via methanolysis and hydrolysis. In *Recycling of polyethylene terephthalate bottles* 85–108. William Andrew Publishing.
Hong, M., & Chen, E. Y. (2017). XChemically recyclable polymers: a circular economy approach to sustainability. *Green Chemistry, 19*, 3692–3706.

Hopewell, J., Dvorak, R. and Kosior, E. (2009). Plastics recycling: challenges and opportunities. *Philosophical Transactions of the Royal Society B: Biological Sciences, 364*, 2115–2126.

https://depositreturnscheme.zerowastescotland.org.uk/

https://docs.european-bioplastics.org/publications/bp/EUBP_BP_Energy_recovery.pdf

https://ec.europa.eu/environment/archives/waste/reporting/pdf/PPW_report_2010_2012.pdf.

https://www.ellenmacarthurfoundation.org/our-work/activities/new-plastics-economy/publications/reuse-rethinking-packaging

https://futurenviro.es/en/el-sistema-de-deposito-devolucion-y-retorno-aleman-sddr/

https://ourworldindata.org/plastic-pollution.

https://wastenarratives.com/2016/05/17/five-rs-of-circular-economy-reduce-reuse-refurbish-repair-and-recycle/.

https://www.circulareconomyasia.org/mechanical-recycling/

https://www.corpseed.com/service/extended-producers-responsibility-epr-authorization.

https://www.ellenmacarthurfoundation.org/assets/downloads/publications/Ellen-MacArthur-Foundation-Towards-the-Circular-Economy-vol.1.pdf.

https://www.exeley.com/exeley/journals/architecture_civil_engineering_environment/12/4/pdf/10.21307_ACEE-2019-055.pdf

https://www.plasticseurope.org/application/files/9915/1708/0036/20170828view_paper_on_mechanical_and_organic_recycling_07072017.pdf.

https://www.wri.org/initiatives/platform-accelerating-circular-economy-pace.

Iaquaniello, G., Centi, G., Salladini, A., Palo, E., & Perathoner, S. (2018). Frontispiece: Waste to Chemicals for a Circular Economy. *Chemistry–A European Journal, 24*, 46.

Ignatyev, I. A., Thielemans, W., & Vander Beke, B. (2014). Recycling of polymers: a review. *ChemSusChem*, 7, 1579–1593.

Jambeck, J. R., Geyer, R., Wilcox, C., Siegler, T. R., Perryman, M., Andrady, A., Narayan, R. and Law, K. L. (2015). Plastic waste inputs from land into the ocean. *Science, 347*, 768–771.

Jambeck J., Hardesty B. D., Brooks A. L., Friend T., Teleki K., Fabres J., Beaudoin Y., Bamba A., Francis J., Ribbink A. J. (2017). Challenges and emerging solutions to the landbased plastic waste issue in Africa. *Marine Policy*, 96, 256–263.

Jones, G. O., Yuen, A., Wojtecki, R. J., Hedrick, J. L., & Garcia, J. M. (2016). Computational and experimental investigations of one-step conversion of poly (carbonate). s into value-added poly (aryl ether sulfone). s. *Proceedings of the National Academy of Sciences, 113*, 7722–7726.

Kim, J. G. (2020). Chemical recycling of poly (bisphenol A carbonate). *Polymer Chemistry, 11*, 4830–4849.

Kovacic, Z., Strand, R., & Völker, T. (2019). *The circular economy in Europe: Critical perspectives on policies and imaginaries*. Routledge.

Kunz, N., Mayers, K., & Van Wassenhove, L. N. (2018). Stakeholder views on extended producer responsibility and the circular economy. *California Management Review, 60*, 45–70.

Layman J. M., Gunnerson M., Schonemann H., Williams K. (2017). Method for purifying contaminated polypropylene (WO2017003796A1)., The Procter & Gamble Company.

Leal Filho, W., Saari, U., Fedoruk, M., Iital, A., Moora, H., Klöga, M., & Voronova, V. (2019). An overview of the problems posed by plastic products and the role of extended producer responsibility in Europe. *Journal of Cleaner Production, 214*, 550–558.

Lifset, R. J. (1993). Take it back: extended producer responsibility as a form of incentive-based environmental policy. *Journal of Resource Management and Technology, 21*, 163–175.

Lindhqvist, T., & Lifset, R. (2003). Can we take the concept of individual producer responsibility from theory to practice?. *Journal of Industrial Ecology, 7*, 3–6.

Liu, Z., Adams, M., Cote, R. P., Chen, Q., Wu, R., Wen, Z., Liu, W. & Dong, L. (2018). How does circular economy respond to greenhouse gas emissions reduction: An analysis of Chinese plastic recycling industries. *Renewable and Sustainable Energy Reviews, 91*, 1162–1169.

MacArthur, D. E., Waughray, D., & Stuchtey, M. R. (2016). The new plastics economy. Rethinking the Future of Plastics.

McNamara D., Murray M. (2011). Conversion of waste plastics material to fuel (WO2011077419A1)., Cynar Plastics Recycling Limited.

Meran, C., Ozturk, O., & Yuksel, M. (2008). Examination of the possibility of recycling and utilizing recycled polyethylene and polypropylene. *Materials and Design, 29*, 701–705.

Merli, R., Preziosi, M. & Acampora, A. (2018). How do scholars approach the circular economy? A systematic literature review. *Journal of Cleaner Production, 178*, 703–722.

Nishida, H. (2011). Development of materials and technologies for control of polymer recycling. *Polymer Journal, 43*, 435–447.

Oliveux, G., Dandy, L. O., & Leeke, G. A. (2015). Degradation of a model epoxy resin by solvolysis routes. *Polymer Degradation and Stability, 118*, 96–103.

Paletta, A., Leal Filho, W., Balogun, A. L., Foschi, E., & Bonoli, A. (2019). Barriers and challenges to plastics valorisation in the context of a circular economy: Case studies from Italy. *Journal of Cleaner Production, 241*, 118149.

Payne, J., McKeown, P., & Jones, M. D. (2019). A circular economy approach to plastic waste. *Polymer Degradation and Stability, 165*, 170–181.

Pedersen, C. S. (2018). The UN sustainable development goals (SDGs). are a great gift to business. *Procedia Cirp, 69*, 21–24.

Peydro, M. A., Parres, F., Crespo, J. E., & Navarro, R. (2013). Recovery of recycled acrylonitrile–butadiene–styrene, through mixing with styrene–ethylene/butylene–styrene. *Journal of Materials Processing Technology, 213*, 1268–1283.

Ragaert, K., Delva, L., & Van Geem, K. (2017). Mechanical and chemical recycling of solid plastic waste. *Waste Management, 69*, 24–58.

Rebeiz, K. S. & Craft, A. P. (1995). Plastic waste management in construction: technological and institutional issues. *Resources, Conservation and Recycling, 15*, 245–257.

Scaffaro, R., Botta, L., & Di Benedetto, G. (2012). Physical properties of virgin-recycled ABS blends: Effect of post-consumer content and of reprocessing cycles. *European Polymer Journal*, 48, 637–648.

Schyns, Z. O., & Shaver, M. P. (2020). Mechanical Recycling of Packaging Plastics: A Review. *Macromolecular Rapid Communications, 42*, 2000415.

Sheldon, R. A., & Norton, M. (2020). Green chemistry and the plastic pollution challenge: towards a circular economy. *Green Chemistry, 22*, 6310–6322.

Siddique, R., Khatib, J., & Kaur, I. (2008). Use of recycled plastic in concrete: A review. *Waste Management, 28*, 1835–1852

Sikdar, S. (2019). Circular economy: Is there anything new in this concept?. *Clean Technologies and Environmental Policy, 21*, 1173–1175.

Simon, B. (2019). What are the most significant aspects of supporting the circular economy in the plastic industry?. *Resources, Conservation and Recycling, 141*, 299–300.

Singh, N., Hui, D., Singh, R., Ahuja, I. P. S., Feo, L., & Fraternali, F. (2017). Recycling of plastic solid waste: A state of art review and future applications. *Composites Part B: Engineering, 115*, 409–422.

Spangenberg, J. H., Fuad-Luke, A., & Blincoe, K. (2010). Design for Sustainability (DfS).: the interface of sustainable production and consumption. *Journal of Cleaner Production, 18*, 1485–1493.

Song, J. H., Murphy, R. J., Narayan, R., & Davies, G. B. H. (2009). Biodegradable and compostable alternatives to conventional plastics. *Philosophical Transactions of the Royal Society B: Biological Sciences, 364*, 2127–2139.

Thompson, R. C., Moore, C. J., Vom Saal, F. S. and Swan, S. H. (2009). Plastics, the environment and human health: current consensus and future trends. *Philosophical Transactions of the Royal Society B: Biological Sciences, 364*, 2153–2166.

Thompson, R. C. (2015). Microplastics in the marine environment: sources, consequences and solutions. In *Marine Anthropogenic Litter*, Springer, Cham., 185–200.

Tournier, V., Topham, C. M., Gilles, A., David, B., Folgoas, C., Moya-Leclair, E., Kamionka, M.-L., Desrousseaux, H., Texier, S., Gavalda, M., Cot, E., Guémard, M., Dalibey, J., Nomme, G., Cioci, S., Barbe, M., Chateau, I., André, S., Duquesne & Marty, A. (2020). An engineered PET depolymerase to break down and recycle plastic bottles. *Nature, 580*, 216–219.

Tukker, A. (2002). Plastics waste: feedstock recycling, chemical recycling and incineration.

Valavanidis, A., Iliopoulos, N., Gotsis, G., & Fiotakis, K. (2008). Persistent free radicals, heavy metals and PAHs generated in particulate soot emissions and residue ash from controlled combustion of common types of plastic. *Journal of Hazardous Materials, 156*, (1-3), 277–284.

Wei, R., and Zimmermann, W. (2017). Microbial enzymes for the recycling of recalcitrant petroleum-based plastics: how far are we? *Microbial Biotechnology 10*, 1308–1322.

Welle, F. (2011). Twenty years of PET bottle to bottle recycling—an overview. *Resources, Conservation and Recycling, 55*, 865–875.

Wright, S. L., & Kelly, F. J. (2017). Plastic and human health: a micro issue?. *Environmental Science and Technology, 51*, 6634–6647.

Yao J. Y., Wang Y. W., Muppaneni T., Shrestha R., Le Roy J., Figuly G. D., Freer E. (2019). Methods for the decomposition of contaminated plastic waste (WO2019204687A1)., Biocellection Inc., USA.

Yoshida, S., Hiraga, K., Takehana, T., Taniguchi, I., Yamaji, H., Maeda, Y., et al (2016). A bacterium that degrades and assimilates poly(ethylene terephthalate). *Science 351*, 1196–1199.

Zheng, J. and Suh, S. (2019). Strategies to reduce the global carbon footprint of plastics. *Nature Climate Change, 9*, 374–378.

2 Additive Manufacturing for Circular Economy of Recycled Plastic

Balwant Singh, Jasgurpreet Singh, and Raman Kumar

2.1 INTRODUCTION TO PLASTIC WASTAGE

Plastics manufacturing rose over 24 times between 1964 and 2015, reaching 322 million tonnes of waste (Mt) in 2015. It is anticipated to almost double in 2035 and nearly quadruple by 2050. Over time, the number of industries throughout the world has increased. As the number of industries grows, so does the amount of garbage produced. As a result, it is critical for current industries to create a long-term waste management strategy. Manufacturers generate a variety of debris (Barra & Leonard, 2018). This can take the form of a solid, a liquid, or a gas. The majority of them had been vetted before being discarded. However, there still is one sort of garbage that is not appropriately managed, namely plastic waste, which is frequently discarded without being treated. It can harm the ecology, particularly the aquatic environment. Individuals discard plastic debris immediately for a range of factors, including the fact that handling plastic waste is complex and time-consuming. Plastic usage is also difficult to mitigate as no resource can substitute plastics in respect to properties and characteristics. Plastic is widely used in most sectors as a raw resource for packing and supplementary materials in product manufacturing (Alauddin et al., 1995).

Many plastic materials include hazardous chemical additions, like persistent organic pollutants (POPs), that have been connected to cancer, psychological, hormonal, and behavioral disabilities. Some plastics are difficult to reuse without causing these compounds to persist. In comparison to other countries, Asians utilize over 30% of global plastic, following by the USA, Europe, and other regions. Throughout 100 metric tons of waste have been used annually in numerous manufacturing sectors across the globe (Stenmarck et al., 2017).

Plastic exists today and will continue to exist in the future. The Plastic Proactive decision acknowledges the importance of plastics in our homes and economy; in fact, it takes efforts not to demonize the product while highlighting the damage done by our failure to effectively handle it.

DOI: 10.1201/9781003184164-2

2.2 RECYCLING OF PLASTICS

The technique of recycling trash or waste material and turning the substances into usable and valuable goods is known as recyclable plastic. The plastic recycling industry is the name for this operation. The purpose of recyclable plastic is to eliminate excessive levels of plastic pollution while reducing the demand for raw resources to make fresh plastic items. This method preserves energy and keeps plastic out of dumps and unanticipated locations like the ocean. The recycling process is critical as a means of dealing with current garbage and as part of a circular economy and negligible systems that wish to minimize waste production and promote sustainable development. The existing waste production and decommissioning behaviors have socioeconomic, ecologic, and economic impacts, and if it is the problem of nano plastics or an approximate $2.5 trillion in destruction and ended up losing reserves to the fishing industry, marine, leisure activities, and worldwide wellness, the effect is undeniable (Germany, 1976).

Addressing the issues created by plastics, on the other hand, is not easy, and there is insufficient knowledge about the disposal of plastic. Ignoring the fact that possible problems were initially identified in the 1960s, there has traditionally been a lot of resistance to serious reform, particularly from the plastics sector. Most users are searching for environmentally friendly materials and learning why the recycling process is vital, so the tide appears to be shifting on this problem. There seems to be a shortage of awareness on how to reuse more products nowadays since both individuals and corporations seek to do so. This causes pollution, either through combining non-recyclable plastics with recyclable materials or by attempting to reuse plastics that have been contaminated by substances like solvents, chemicals, or food leftovers, which slows down the reprocessing even more. Each of these issues can result in plastics being discarded instead of reused (Sardon & Li, 2020).

An additional problem is that the objects themselves are complicated. While a few items, such as plastic cups and other tetra packs, are typically composed of a single standard plastic, making them easy to recover, so many are intended to employ a combination of polymers, which can present major problems in our existing plastic reprocessing. Furthermore, many objects are made up of a combination of plastic and non-plastic materials such as metal or wood. Unfortunately, these items will not even be taken to a recycling center (Cooper, 2013).

2.3 PROCESSING STEPS FOR PLASTICS RECYCLING

The recycling process has advanced significantly in recent years, and it can now be broken down into six simple phrases.

1. **Collection and Distribution:** The collecting of post-consumer waste from households, companies, and organizations is the preliminary stage in mechanical reprocessing. Local authorities or corporate entities can accomplish this, with the latter being a popular solution for organizations.

Bringing materials to common collection places, like dedicated bins or stations, is also another solution. This could be as basic as a container collection on a corner of the street or as complicated as disposal of solid waste dump with huge areas for diverse recyclable and non-recyclable materials (Shen & Worrell, 2014).

2. **Sorting and Categorizing:** Filtering is the second stage in plastic reprocessing. There are various forms of plastics that waste pickers must separate from one another. Plastics can also be classified based on their colors, density, and intended usage. This is performed in the recycling center by machinery and is a key step in increasing plant productivity and avoiding pollution of end goods (Engineering et al., 2006).
3. **Washing:** Cleansing is an important stage in plastic reprocessing because it eliminates contaminants that might stymie the operation or even damage a stock of recyclable materials. Contaminants like labeling requirements and adhesive, and also dust and foodstuff waste, are usually directed in this stage. Although plastic is frequently cleaned at this step, it is vital to note that this is still necessary to ensure that plastics are as clean as possible before destruction and retrieval (Santos et al., 2005).
4. **Shredding:** After that, the plastic is sent into crushers, which break it into considerably thin strips. Despite molded plastic items, such tiny bits can be treated for recycling in subsequent phases. The smaller plastic fragments can also be used in varied uses without any more treatment, such as that of an asphalt ingredient or easily sold as raw goods.

   Tearing down the plastic into tiny pieces also makes it easier to find any leftover contaminants. This seems to be certainly relevant to pollutants like metal, which might not be eliminated by cleaning but may now be easily extracted by magnetism (Tsuchida et al., 2009).
5. **Plastics Identification and Separation:** The grade and durability of the plastic parts are assessed here. They are only separated according to size and distribution, which is determined by floating plastic particles in a container. Following that, a study for "air categorization," which assesses the size of the plastic fragments, is performed. The shredded plastic is placed in a test section, with finer fragments floating and bigger portions remaining at the base (Ruj et al., 2015).
6. **Extrusion and Compounding:** The recycling of plastic granules is turned into a viable product for makers in this final stage of plastic reprocessing. Pellets are formed by melting and crushing disposal and recycling. Because it is not usually feasible to combine all kinds, classifications, and quality of plastic in a particular plant, various grades of plastic are sometimes shipped to other recycling centers in this last stage (Sohn et al., 2019).

2.4 TYPES OF PLASTICS

There are many different forms of plastic, and there are seven different types to consider when learning about plastic reprocessing and avoiding pollution. These symbols are seen on products, and while it appears to be a "recycling symbol," it denotes the resin type, with some denoting non-recyclable materials.

1. **Polyethylene Terephthalate (PETE):** This resin is being used to make necessities like food cartons and plastic containers for water or soft beverages, and it is among the most prevalent forms of plastic (Westerhoff et al., 2008).
2. **High-Density Polyethylene (HDPE):** This form of plastic, which is stiffer than polyethylene terephthalate and used in things that look to be "stiffer," such as soap containers, beverages containers, soda cans, heavier grocery bags, and non-single-use plastic products including sports, hats, and pipework. This form of plastic can also be recycled (Sogancioglu et al., 2017b).
3. **Polyvinyl Chloride (PVC):** PVC is among the most flexible and widely utilized plastics, with uses including water and waste pipelines (because of its biochemical and physiological resistance), floor, furnishings, and much more. While alternative techniques for recycling PVC have been established, it is not widely used and is rarely met in the conventional plastic collection. This is due in significant part to PVC's toxicity when treated (Turner & Filella, 2021).
4. **Low-Density Polyethylene (LDPE):** While not as robust as HDPE, this low-density plastic is quite durable and is utilized in a variety of goods including bottles, play equipment, and garbage bags. Although this polymer kind is reusable, numerous goods (such as plastic bottles) can be rejected because they offer a danger of blocking equipment and are regarded unworthy of recycling (Kumi-Larbi et al., 2018).
5. **Polypropylene (PP):** This plastic is widely in use in the injection molding process and it can be encountered in everything from plastic containers to medical instruments and apparel. While PP is reusable, it is frequently refused by processing plants considering the problems it causes, resulting in less percentage of recycling than other plastics (Hisham A. Maddah, 2016).
6. **Polystyrene (PS):** This plastic is commonly used in food-safe carton boxes, insulating cartons, and plastic packaging. PS is infrequently reused, although its availability is because it is not an expense (its most prevalent form, expanded polystyrene, is 95% air) and consumes more power to recover than it costs (Sogancioglu et al., 2017a).
7. **Other:** Almost everything fits that description, usually involving a mix of the previous six and therefore less commonly used plastics. Non-petrochemical plastics, such as novel plastic products, polymer, and bioplastics, are also included in this category. As a result, anything having a variety 7 on it is usually not recyclable, and it may have other disposal options.

Currently, the plastics reusing confronts numerous obstacles, however, unlike glass and aluminum, plastics are not fully recyclable, indicating that the recovered product declines and becomes a worse grade with each consecutive treatment.

But, do not overlook the bigger picture. Today's modern plastic recycling industry is way ahead of what it was just a few decades ago, with recycling rates steadily increasing and continuing to do so. Chemical recycling is being developed to keep fewer plastics in the recycling loop for longer. Furthermore, a rising number of plastic-free options are becoming available.

As households and corporations become more involved in the plastic recycling process, we may expect it to increase. This, including the transition to innovative products and modern substitutes, represents modest but steady progress in the correct direction.

As it becomes more conscious of how to utilize assets and make goods, the recyclers' method is intended to see a massive benefit. While waste disposal ideas, including the circular economy and waste minimization, are generally associated with a relocation away from single-use plastics, reuse will continue to be a part of the waste management strategy for the coming years, enabling us to continue moving away from useless plastics while increasing the recycled content of those that persist necessary (Alauddin et al., 1995).

2.5 CIRCULAR ECONOMY

The circular economy is a type of economy in which resources are intended to be utilized rather than discarded. Products and the processes are part of it and designed from the start to assure that no materials are wasted, no pollutants are released, and that every procedure, resource, and element is used to its full potential. The circular economy, when properly implemented, helps the community, the ecology, and the economic system. All packing must be adjusted to match into a process, whether that structure is for reusing, recycling, or composting. The circular economy for plastics has been shown in Figure 2.1.

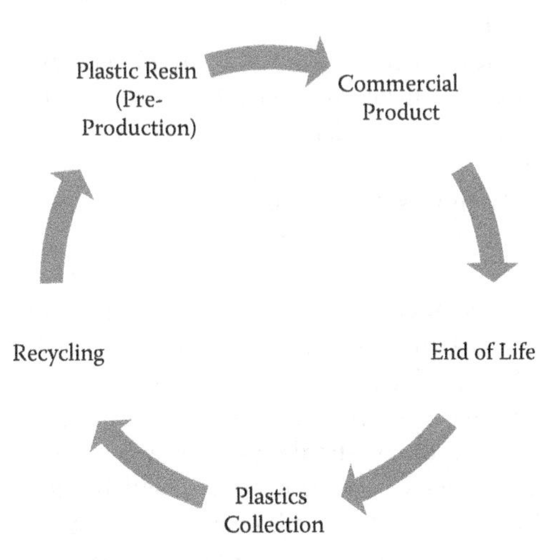

FIGURE 2.1 Circular economy of plastics.

There are numerous advantages to using plastic. At the very same time, certain harmful commodities on the marketplace must be removed to establish a circular economy, and packaging material can occasionally be omitted entirely while preserving usability.

While enhancing recycling is critical, it will not be able to reuse our own out of another present plastic crisis. Wherever possible, recycling marketing strategies (also known as an "inner loop" in the circular economy) must be investigated as a preferred alternative for minimizing the demand for single-use plastic packaging. Reuse methods that provide a financially viable option for at least 20% of plastic packaging must be applied in practice and on a large scale (Sempels & Hoffmann, 2013).

This will necessitate a range of business model creativity and transformation, as well as package design and reprocessing technology. When an adequate gathering and compost system is required but mostly not in place, recyclable plastic wrapping is not a universal solution, but rather one for particular, specialized uses. Maintain all of the pieces of plastic that are use in circulation to keeping them out from the environment and out of the economy. There should be no plastic in the ecosystem. Landfills, burning, and waste-to-energy are not protracted circular economy possibilities. Governments are critical in establishing accurate sampling facilities, promoting the creation of associated identity funding sources, and creating an appropriate legal and policy environment. Further than the construction and usage of its packaging, companies manufacturing and/or selling it have an obligation to contribute to it being gathered and recycled, repurposed, or composted in reality (Korhonen et al., 2018).

There are six main points in the concept for a circular economy for plastic:

1. Plastic packaging that is troublesome or superfluous must be eliminated through redesigning, creativity, and alternative delivery mechanisms.
2. Wherever possible, reuse models are used to reduce the requirement for single-use packaging.
3. Every piece of plastic packaging is 100% reused, reusable, or biodegradable.
4. In reality, all packing material is recycled, repurposed, or composted.
5. The use of plastic is completely unrelated to the depletion of scarce resources.
6. All plastic packaging is free of dangerous substances, and all people's health, safety, and rights are protected.

Plastic does not become trash or contamination in the new plastics economy. To realize this goal and build a circular economy for plastic, three activities are required. Remove all hazardous and superfluous plastic things from your home. Develop new ways to make the plastics we do need recyclable, reusable, or biodegradable. Maintain all of the plastic things we use in circulation to keep them out of the environment and out of the economy. It will be impossible to achieve a circular economy for plastic without eliminating it. With consumption of plastic packaging expected to get double in the next two decades, keeping this ever-increasing flow of plastics out of the economy and out of the ecosystem would be difficult. We need to limit the amount of material that needs to be circulated to establish a circular economy.

The term "material circulation" refers to maintaining the materials used to make packaging in circulation throughout the economy. This is accomplished by creating a specialized system that includes gathering and filtering, a physical-chemical or biological breakdown process, and then the rebuilding of a reintroduced substance. Material circulation occurs when packing materials re-enter the economy in packaging applications; however, it is not when packaging materials are converted into roadways (Ellen MacArthur Foundation, 2013).

2.6 FUTURE CHALLENGES AND SCOPE OF THE CIRCULAR ECONOMY

A circular economy reimagines existing consumption habits in such a way that organization and growth promote strong economic, socioeconomic, and ecologic outcomes across supply chains, marketing strategies, and life cycles, from raw material selection to product/service design to recycling and end-of-life management. The recent global market model is based on the availability of low-cost, widely accessible commodities. The waste that is accumulating in our ocean resources is a clear sign of this take-make-dispose economy, also known as the linear economy. Single-use packaging materials pollution has become a serious issue in recent years, accounting for over half of all plastic garbage generated globally in 2015. Single-use plastic packaging is relatively light, inexpensive, versatile, and seems to last for hundreds of years, seem to be the same features that make it useful in keeping goods and being mass-produced. However, because of its minimal value, it is difficult to dispose of and resale, making it more vulnerable to misuse and improperly disposed of. The coronavirus illness (COVID-19) crisis has also put the global financial system and daily lives on hold, and it has also disturbed the waste supply chain, resulting in a huge increase in healthcare and plastics trash. This is a chance to reconstruct to use more systematic solutions, develop environmentally friendly and robust lifestyles, and enhance technology advancement. Companies, governments, and individuals may benefit from technical advancements to assist them to move to a more equitable circular economy more rapidly and easily.

Trash plastics and their environmental consequences have gotten a lot of attention recently, from China's prohibition on plastic waste importation to the EU's efforts to drastically cut plastic use. When considering prospective remedies, including scientific information transmission or commercial ramping up initiatives, it is becoming clear that clean-ups alone are insufficient, and that system innovation, such as the circular economy, is required (Payne et al., 2019).

2.7 CASE STUDY

In the present experimental work, the ABS plastic was taken from toys and keyboard keys from the domestic waste that was recycled and 3D printing was done. The different testing was also done to analyze the different properties according to the input parameters. The flow diagram of the experimental work is shown in Figure 2.2.

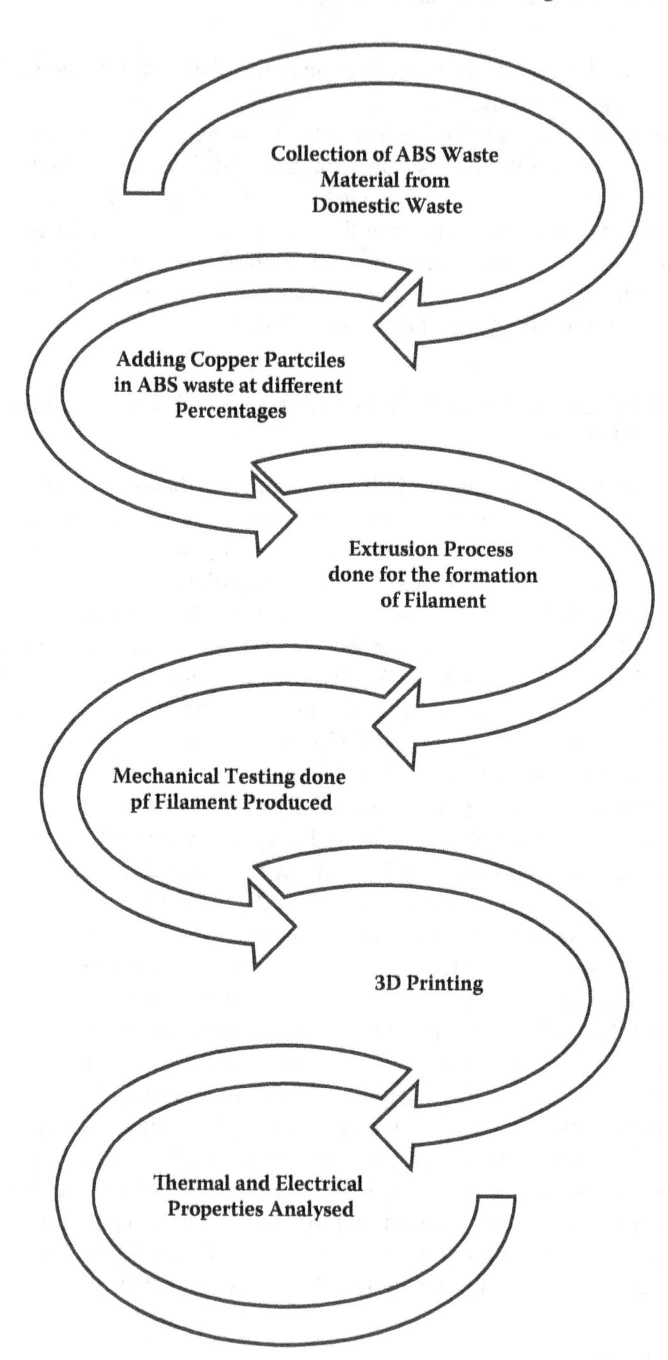

FIGURE 2.2 Flow chart of experimental work.

First of all, the waste was collected from the domestic waste in which toys and keyboard keys are included, made up of ABS material. After the collection of ABS waste material, it was mixed with copper particles. The copper particles used in the present work were of three different sizes (100 Mesh, 200 Mesh, and 400 Mesh), and also the variable percentage of copper particles from 1% to 10% weight in waste has been added to check the different properties of the composite material. The melt flow rate (MFR) was analyzed for the different proportions of copper particles and ABS waste material. In the present work, first of all, single-particle size composition was done by using the weighing machine in which the composite materials were obtained such as waste ABS with 100 Mesh copper particles, waste ABS with 200 Mesh copper particles, and waste ABS with 400 Mesh copper particles. Secondly, double particle size composition was mixed in weighing machines such as waste ABS with 100 Mesh and 200 Mesh copper particles, waste ABS with 100 Mesh and 400 Mesh copper particles, and waste ABS with 400 Mesh and 200 Mesh copper particles. Thirdly, triple particle size compositions were manufactured with a weighing machine in which waste ABS with 100 Mesh, 200 Mesh, and 400 Mesh copper particles.

This method is one of the most common methods used by the different industries when describing the exact composition of the material needed according to their melt flow rate for the specific FDM machine is defined.

2.7.1 Melt Flow Testing of ABS with Single, Double, and Triple Particle Size of Copper Particles

The waste ABS material was mixed with copper particles at different proportions to check the flow rate at every stage by using the 2.180 kg ram weight and 220°C temperature of the melt flow tester. In single-size copper particles with waste ABS, the melt flow rate has been described in Table 2.1 and Figure 2.3.

The double-size copper particles with waste ABS have been described in Table 2.2 and Figure 2.4.

Lastly, the melt flow index testing was done for the waste ABS with triple-particle size of copper has been shown in Table 2.3 and Figure 2.5.

Overall, the best fluidity 2.29 gm/10 min was obtained while performing the melt flow test at 10% copper particles and 90% ABS, which is best suitable for making the feedstock filament for 3D printing in a screw extruder, and the best tensile strength was obtained at 8% to 10% copper particles in ABS at 230°C. ABS material was employed in a mixture with copper particles for further investigation.

2.7.2 Manufacturing of Feedstock Filament for 3D Printing with Twin-Screw Extruder Machine

A twin-screw extruder was used to produce wire with a diameter of 1.75 mm in this study. The twin-screw extruder machine can rotate at various speeds and print at various temperatures. The rotating speed was set at 85 revolutions per minute, and the printing temperature of the nozzle was fixed at 230°C in this study. Then, using the hopper on top of the screw extruder, slowly pour in all of the mixed material so that it can melt correctly in the machine and the wire can be created properly.

TABLE 2.1
MFI of waste ABS reinforced with single-particle size copper

Material Composition	1st Reading	2nd Reading	3rd Reading	Mean	SD
100 Mesh Copper Particles with Waste ABS (gm/10 min)					
1% Copper & 99% ABS	2.16	2.23	2.31	2.23	0.074
2.5% Copper & 97.5% ABS	1.55	1.77	1.67	1.66	0.100
5% Copper & 95% ABS	1.46	1.60	1.70	1.58	0.121
200 Mesh Copper Particles with waste ABS (gm/10 min)					
1% Copper & 99% ABS	1.70	1.78	2.12	1.87	0.223
2.5% Copper & 97.5% ABS	2.47	2.16	2.01	2.21	0.22
5% Copper & 95% ABS	2.63	2.34	2.37	2.43	0.157
8% Copper & 92%ABS	2.37	2.44	2.30	2.37	0.068
10% Copper & 90% ABS	2.27	2.35	2.21	2.27	0.071
400 Mesh Copper Particles with waste ABS (gm/10 min)					
1% Copper & 99% ABS	1.46	1.62	1.64	1.57	0.097
2.5% Copper & 97.5% ABS	1.56	1.74	1.67	1.66	0.089
5% Copper & 95% ABS	2.09	1.70	1.90	1.90	0.199

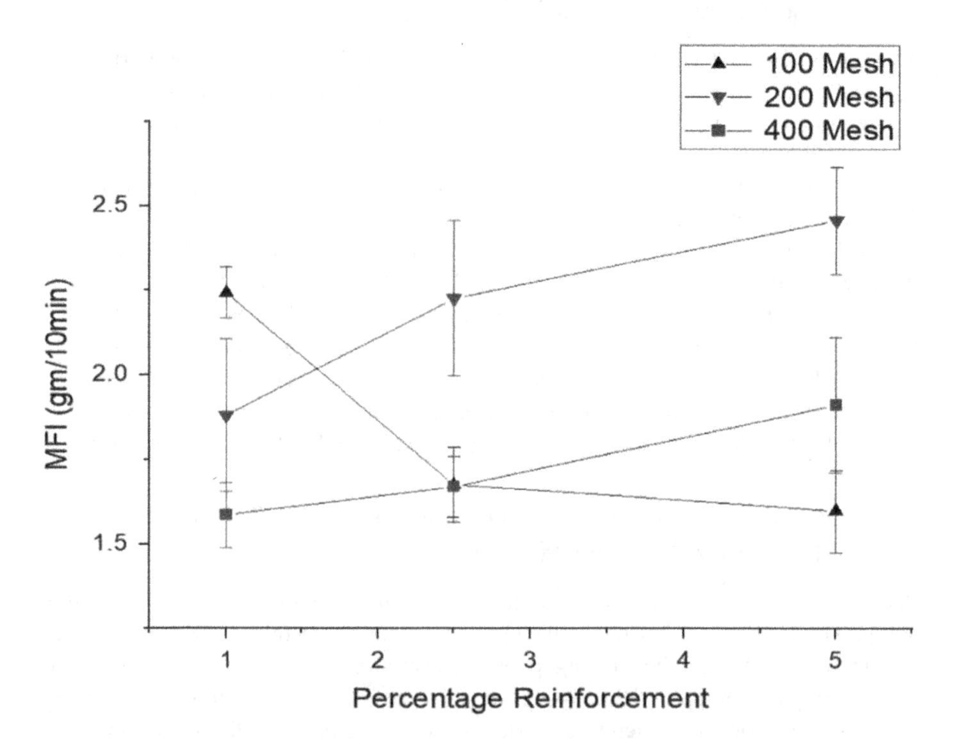

FIGURE 2.3 Melt flow rate of single-size copper particles with ABS.

TABLE 2.2
MFI of waste ABS reinforced with double-particle size copper

Material Composition	1st Reading	2nd Reading	3rd Reading	Mean	SD
100 Mesh + 200 Mesh Copper Particles with ABS					
1% Copper & 99% ABS	2.42	2.38	2.39	2.40	0.022
3% Copper & 97% ABS	2.30	2.30	2.37	2.32	0.031
6% Copper & 94% ABS	2.22	2.30	2.24	2.26	0.039
200 Mesh + 400 Mesh Copper Particles with ABS					
1% Copper & 99% ABS	2.05	1.87	2.01	1.97	0.097
3% Copper & 97% ABS	1.66	1.76	1.63	1.69	0.064
6% Copper & 94% ABS	2.25	2.22	2.22	2.22	0.011
100 Mesh + 400 Mesh Copper Particles with ABS					
1% Copper & 99% ABS	1.91	1.70	2.29	1.97	0.27
3% Copper & 97% ABS	2.07	2.06	1.80	1.98	0.149
6% Copper & 94% ABS	2.24	2.35	2.27	2.29	0.055

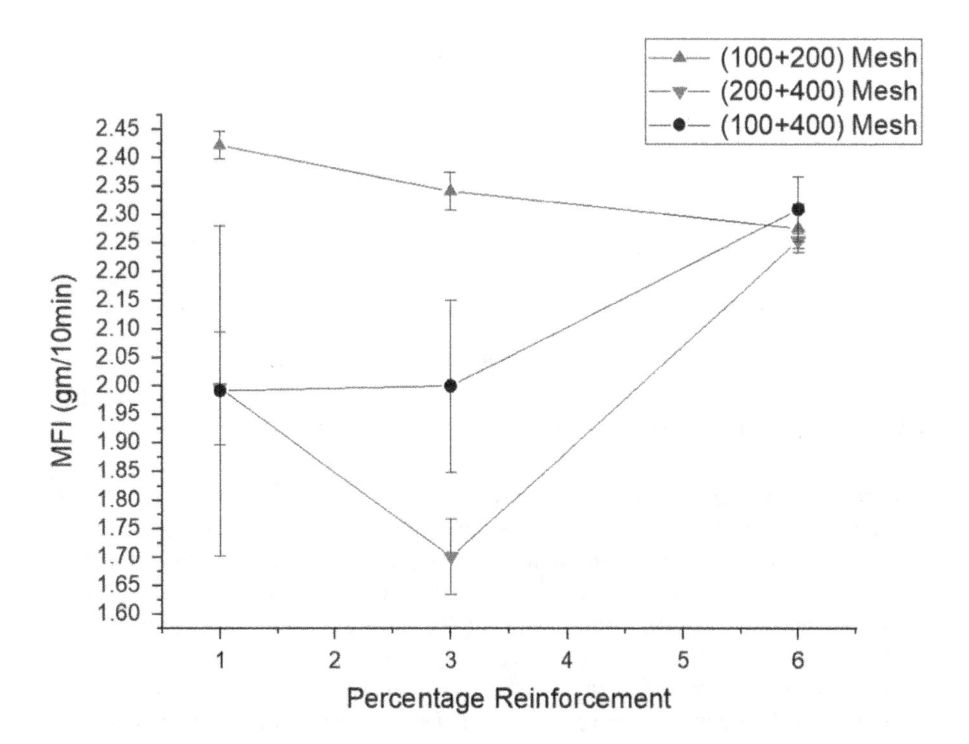

FIGURE 2.4 Melt flow rate of double-size copper particles with ABS.

TABLE 2.3
MFI of waste ABS reinforced with double-particle size copper

A mixture of 100 Mesh, 200 Mesh, and 400 Mesh Copper Particles with ABS					
3% Copper & 97% ABS	1.77	1.85	1.99	1.87	0.10
6% Copper & 94% ABS	2.22	2.30	2.26	2.26	0.030
8% Copper & 92% ABS	2.62	2.32	2.36	2.43	0.157

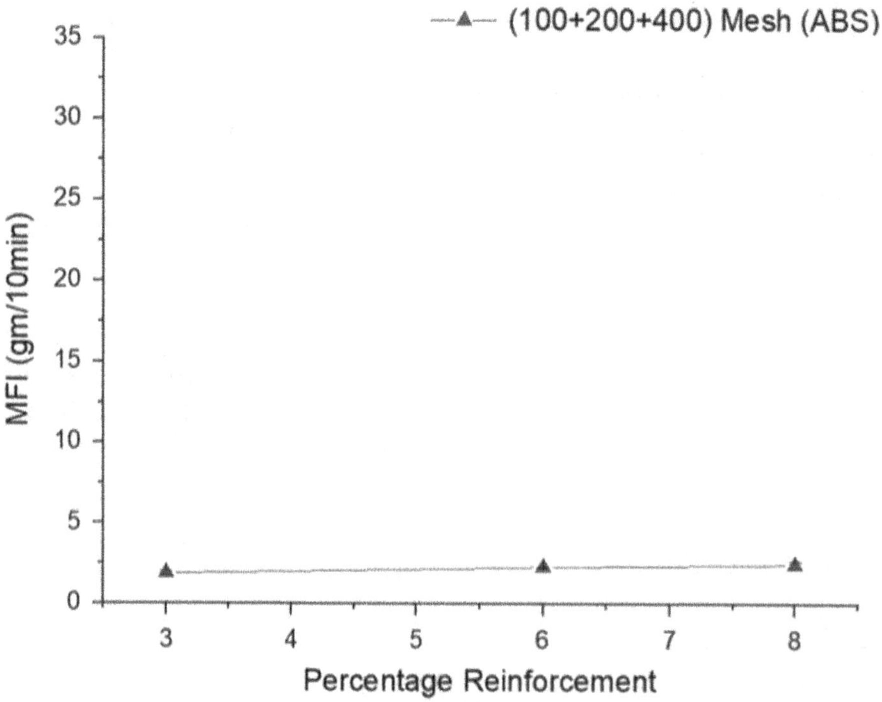

FIGURE 2.5 Melt flow rate of triple-size copper particles with ABS.

One reel was placed in front of the screw extruder to make the wire, which was rotated at an appropriate speed by the stepper motor in it, resulting in wire diameter of 1.75 mm. The twin-screw extrusion machine, produced by Thermo Scientific Pvt. Ltd., Europe, is shown in Figure 2.6. Process 16 parallel twin-screw extruders were the machine's model. The details of the twin-extruder machine are shown in Table 2.4.

Three factors were varied in this study, which was chosen using the Taguchi Method (L9 Orthogonal Array) and Mini Tab software, while the rest of the parameters were fixed. Using the Mini Tab software with an L9 Orthogonal Array Taguchi Design, the three parameters, printing temperature, layer height, and infill

FIGURE 2.6 Twin-extruder machine.

TABLE 2.4
Details of twin-extruder machine

Design Model	Max. Screw Speed (rpm)	Temperature Zones	Operating Concept	Drive Power (KW)
Counter-Rotating	360	2	PC	0.4

pattern, were continually varied into three different levels, and the table was generated with multiple combinations. Printing temperature (230°C, 240°C, 250°C), layer height (0.1 mm, 0.15 mm, 0.2 mm), and infill pattern were all chosen from a variety of options (lines, zig-zag, triangles). The Taguchi method experiment technique was used to test nine different combinations of printing temperature, layer height, and infill patterns. It is made up of three discrete process variables that must be divided into three different values.

2.8 THERMAL AND ELECTRICAL PROPERTIES ANALYZED

The electrical and thermal conductivity of various ABS and copper particle compositions were investigated (200 Mesh). The "Megger" equipment was used to assess the resistance of the composite material, from which the conductivity value could be easily estimated utilizing the relationship between resistivity and conductivity, whereas the Lee's Disc Method was utilized to analyze the thermal conductivity. It was assumed that while increasing the copper particles in waste ABS the thermal and electrical conductivity was improved. The graphs for different compositions have been shown in Figure 2.7 and Figure 2.8.

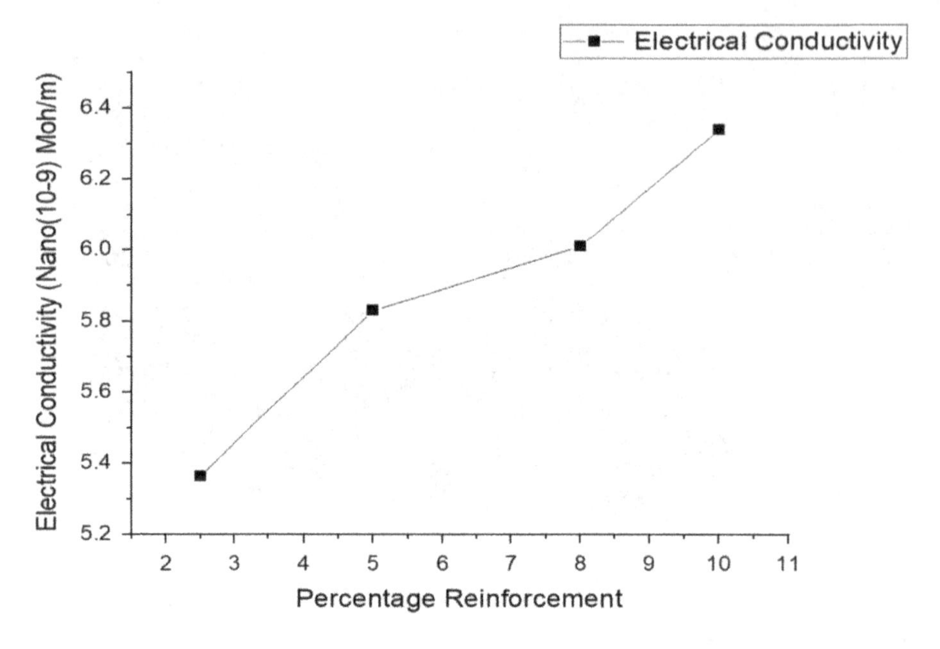

FIGURE 2.7 3D printing machine.

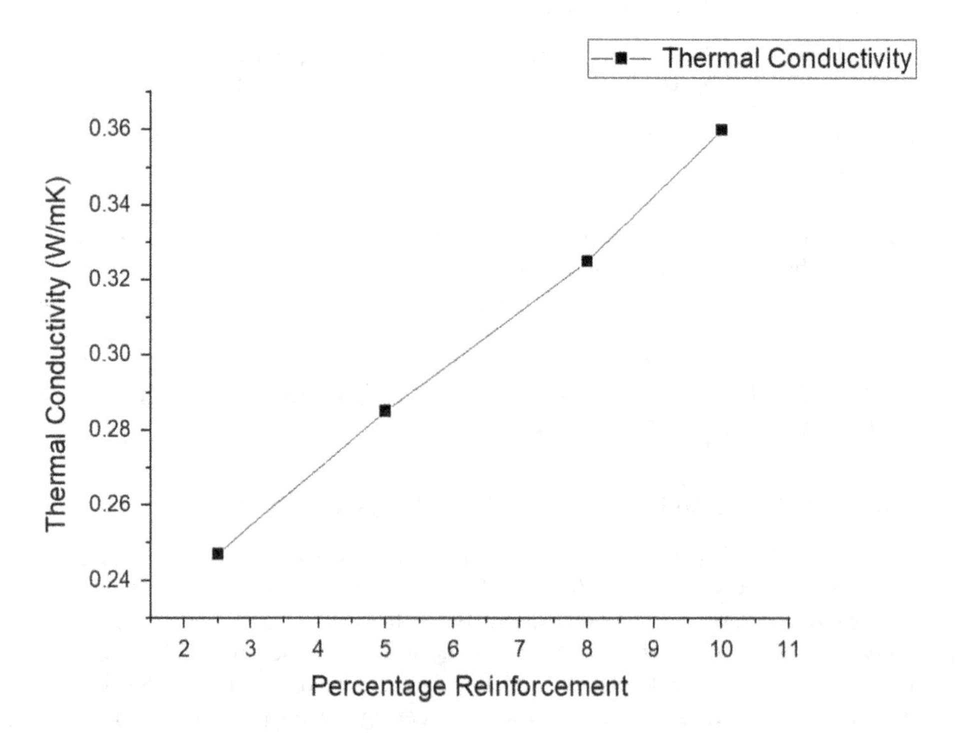

FIGURE 2.8 Graph in between electrical conductivity as per increase of copper particles in waste ABS.

2.9 THERMAL AND ELECTRICAL PROPERTIES ANALYZED

In the present work, the waste ABS plastic material was recycled in addition with copper particles at different percentages from 1% to 10% to improve the thermal and electrical conductivity of the final product manufactured by FDM (fused deposition modeling) process. The melt flow index testing was done for different compositions that vary from 1% to 10% copper particle with recycled ABS plastic. The best fluidity, 2.29 gm/10 min, was obtained while performing the melt flow test at 10% copper particles and 90% ABS, which is best suitable for making the feedstock filament for 3D printing in a screw extruder, and the best tensile strength was obtained at 8% to 10% copper particles in ABS at 230°C. Also, it has been analyzed that the thermal and electrical conductivity was improved 113.04% and 199.85%, respectively, by using 200 Mesh copper particles in waste ABS material after recycling.

REFERENCES

Alauddin, M., Choudhury, I. A., El Baradie, M. A., & Hashmi, M. S. J. (1995). Plastics and their machining: A review. *Journal of Materials Processing Technology*, *54*(1–4), 40–46. 10.1016/0924-0136(95)01917-0

Barra, Ricardo & Leonard, S. A. (2018). *Plastics and the circular economy. Scientific and Technical Advisory Panel to the Global Environment Facility.* (June). Retrieved from https://stapgef.org/sites/default/files/publications/PLASTICS for posting.pdf

Cooper, T. A. (2013). Developments in plastic materials and recycling systems for packaging food, beverages and other fast-moving consumer goods. In *Trends in Packaging of Food, Beverages and Other Fast-Moving Consumer Goods (FMCG).* 10.1533/978085 7098979.58

Ellen MacArthur Foundation. (2013). *Towards the circular economy. Journal of Industrial Ecology*, 2, 23–44.

Engineering, F., Bangi, U. K. M., & De, S. (2006). *Development of a Prototype Automated Sorting System for Plastic Recycling D. A. Wahab, A. Hussain, E. Scavino, M. M. Mustafa and H. Basri. 3*(7), 1924–1928.

Germany, W. (1976). *Recycling of Plastics W. Kaminsky, J. Menzel and H. Sinn. 1*, 91–110.

Hisham A. Maddah. (2016). Polypropylene as a Promising Plastic: A Review. *American Journal of Polymer Science.* 10.5923/j.ajps.20160601.01

Korhonen, J., Honkasalo, A., & Seppälä, J. (2018). Circular Economy: The Concept and its Limitations. *Ecological Economics*, *143*, 37–46. 10.1016/j.ecolecon.2017.06.041

Kumi-Larbi, A., Yunana, D., Kamsouloum, P., Webster, M., Wilson, D. C., & Cheeseman, C. (2018). Recycling waste plastics in developing countries: Use of low-density polyethylene water sachets to form plastic bonded sand blocks. *Waste Management*, *80*, 112–118. 10.1016/j.wasman.2018.09.003

Payne, J., McKeown, P., & Jones, M. D. (2019). A circular economy approach to plastic waste. *Polymer Degradation and Stability*, *165*, 170–181. 10.1016/j.polymdegradstab.2019.05.014

Ruj, B., Pandey, V., Jash, P., & Srivastava, V. K. (2015). Sorting of plastic waste for effective recycling. *Int. Journal of Applied Sciences and Engineering Research*, *4*(4), 564–571. Retrieved from www.ijaser.com

Santos, A. S. F., Teixeira, B. A. N., Agnelli, J. A. M., & Manrich, S. (2005). Characterization of effluents through a typical plastic recycling process: An evaluation of cleaning performance and environmental pollution. *Resources, Conservation and Recycling*, *45*(2), 159–171. 10.1016/j.resconrec.2005.01.011

Sardon, H., & Li, Z. C. (2020). Introduction to plastics in a circular economy. *Polymer Chemistry, 11*(30), 4828–4829. 10.1039/d0py90117b

Sempels, C., & Hoffmann, J. (2013). Circular Economy. *Sustainable Innovation Strategy*, 6–9. 10.1057/9781137352613.0008

Shen, L., & Worrell, E. (2014). Plastic Recycling. In *Handbook of Recycling: State-of-the-art for Practitioners, Analysts, and Scientists*. 10.1016/B978-0-12-396459-5.00013-1

Sogancioglu, M., Yel, E., & Ahmetli, G. (2017a). Investigation of the Effect of Polystyrene (PS) Waste Washing Process and Pyrolysis Temperature on (PS) Pyrolysis Product Quality. *Energy Procedia, 118*, 189–194. 10.1016/j.egypro.2017.07.029

Sogancioglu, M., Yel, E., & Ahmetli, G. (2017b). Pyrolysis of waste high density polyethylene (HDPE) and low density polyethylene (LDPE) plastics and production of epoxy composites with their pyrolysis chars. *Journal of Cleaner Production, 165*, 369–381. 10.1016/j.jclepro.2017.07.157

Sohn, J. S., Ryu, Y., Yun, C. S., Zhu, K., & Cha, S. W. (2019). Extrusion compounding process for the development of eco-friendly SCG/PP composite pellets. *Sustainability (Switzerland), 11*(6). 10.3390/su11061720

Stenmarck, Å., Belleza, E. L., Fråne, A., Busch, N., Larsen, Å., & Wahlström, M. (2017). *Hazardous substances in plastics-ways to increase recycling – In cooperation with IVL Svenska Miljöinstitutet AB, CRI*, SINTEF**, VTT****. Retrieved from www.ivl.se

Tsuchida, A., Kawazumi, H., Kazuyoshi, A., & Yasuo, T. (2009). Identification of shredded plastics in milliseconds using Raman spectroscopy for recycling. *Proceedings of IEEE Sensors*, 1473–1476. 10.1109/ICSENS.2009.5398454

Turner, A., & Filella, M. (2021). Polyvinyl chloride in consumer and environmental plastics, with a particular focus on metal-based additives. *Environmental Science: Processes & Impacts*. 10.1039/d1em00213a

Westerhoff, P., Prapaipong, P., Shock, E., & Hillaireau, A. (2008). Antimony leaching from polyethylene terephthalate (PET) plastic used for bottled drinking water. *Water Research, 42*(3), 551–556. 10.1016/j.watres.2007.07.048

3 Hybrid Mechanical and Chemical Recycling of Plastics

Ravinder Sharma, Rupinder Singh, Ajay Batish, and Nishant Ranjan

3.1 INTRODUCTION

Nowadays, thermoplastic polymers have been widely used in the fields of biomedical science, electronic appliances, energy harvesting applications, automobile industries, etc. The use of thermoplastics has increased twentyfold in the last 50 years. Moreover, it is supposed to be double in the next two decades (Moore, 2008). Despite numerous socio-economic applications, the low rate of reusability of plastics is a main concern. For instance, in Australia, every year 95% of total produced plastic was disposed as waste after one-time use (Jubinville et al., 2020). This accumulated plastic waste is mostly burned, which leads to polluting the environment (Verma et al., 2016). All of this is due to the immature recycling/reprocessing technologies. Thus, to attenuate climate change and conserve the pollution-free atmosphere, we need a "3Rs" (reduce, reuse, and recycle) initiative from the concept of "circular economy" worldwide (Grossule, 2019). Thus, for minimal waste of plastics, various newly invented technologies have been used in modern industries (Shekdar, 2009). 3D printing also offers a new path towards the circular economy by reutilization of the post-consumer plastics for fabrication of new products. These materials can be recycled by mechanical or chemical methods for the additive manufacturing process (Depalma et al., 2020; Shanmugam et al., 2020).

The additive manufacturing (AM)/3D printing techniques have attention from various disciplines as a zero-waste-based production process (Ngo et al., 2018). The classification of the AM process is mainly on the basis of liquid, solid, and powder-type raw material (Gardan, 2016). In most commonly used 3D printing processes, semi-molten raw material has been deposited in the form of layers over previously deposited material and the process remains the same until completion of parts (Sharma et al., 2018; Ramya and Vanapalli, 2016). The AM process is a completely controlled process, in which initially a program has been saved in the form of G-code that gives commands to a 3D printer for fabrication of parts [Cummings et al., 2017). Earlier, only thermoplastic-based polymers were used in the 3D printing process; however, nowadays 3D printing of metals, smart materials, and

DOI: 10.1201/9781003184164-3

composites of polymers and metal powder have also been processed with AM technologies (Lee et al., 2017; Mangat et al., 2018; Sharma et al., 2019).

Fused deposition modelling (FDM) has been considered the most widely used 3D printing technoiogy. Thermoplastic polymers or composites based on these polymers have been used in the form of feedstock filament to fabricate a final product (Dickson et al., 2020; Park & Fu, 2021; Singh et al., 2018). The low product development cost and easse of use of FDM printers enhances their applicability from school experiments to industrial production (Singh et al., 2020). Commercial FDM printers provide good control over their process variables such as infill density (ID), infill angle (IA), infill speed (IS), raster width, printing temperature, etc. (Deshwal et al., 2020; Sharma et al., 2020). These process parameters put promising effects in the properties of outcome products (Chacón et al., 2019; Singh et al., 2021; Sharma et al., 2020). It has been reported in the literature that the mechanical strength of the 3D-printed parts reduced with an increase of IA, whereas product strength is directly proportional to the ID (Sharma et al., 2021).

3.1.1 4D Printing

Four-dimensional printing is one step ahead of the 3D printing process, in which time is considered a fourth dimension. 4D printing incorporates the same 3D printing techniques for fabrication of functional prototypes; however, these 3D-printed parts transformed their shapes or other physical properties over time (Haleem et al., 2021; Martinez-Marquez et al., 2018). The materials used for 4D printing are also known as smart materials. These materials transform their shapes in a pre-programmed manner when they come in contact with heat, humidity, electric field, magnetic field, mechanical forces, etc. (Zafar & Zhao, 2020; Sharma et al., 2020). 4D printing is mainly classified into the two following types:

Autonomous 4D printing: In this type of 4D printing, there is no need for any external intervention for transformation in the material. These types of materials react automatically with the change in atmospheric temperature, humidity, etc.

Non-autonomous 4D printing: In non-autonomous 4D printing, an external trigger is required to charge the material for the transformation process. For example, by applying an electric field, magnetic field, or mechanical force, etc., the materials will alter their properties over time.

Thus, integration of smart polymers at the starting stage of 3D printing for fabrication of structures is also known as 4D printing. Smart materials are one of the most attractive research areas in academic and industrial research (Melly et al., 2020; Bengisu and Ferrara, 2018; Sharma et al., 2020).

Shape memory polymers (SMPs), electroactive polymers (EAPs), and other smart polymers are able to change/alter their dimensional or physical properties over some period of time when subjected to some external stimuli. EAP or piezoelectric polymers produce electricity when under mechanical stress, or vice versa of it is known as reverse piezoelectricity. EAPs are generally used in sensor applications and can be easily molded into any shape with integration of AM (Maurya et al., 2018; Sharma et al., 2021; Safaei et al., 2019).

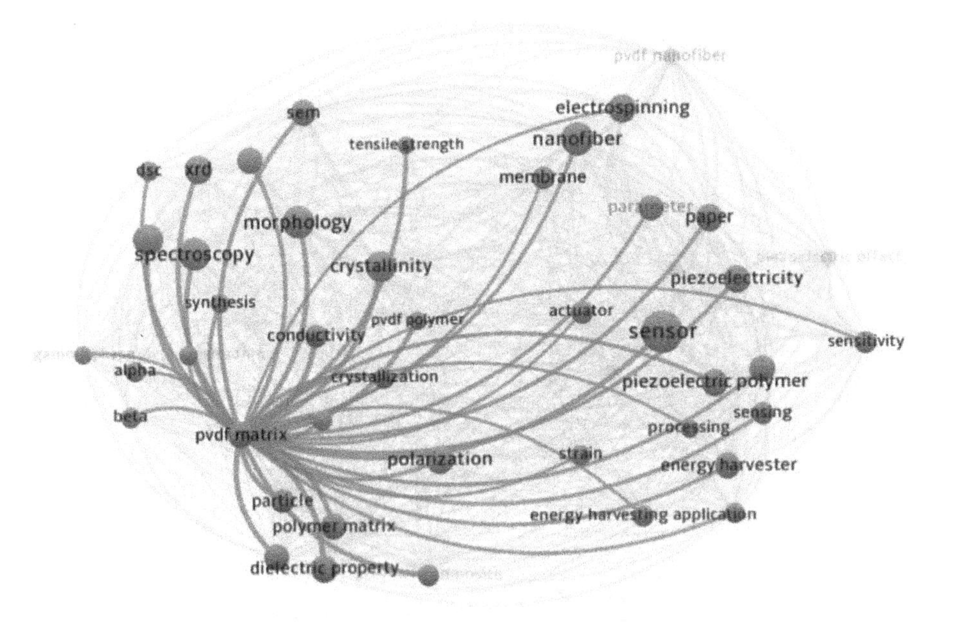

FIGURE 3.1 Bibliographic analysis of PVDF polymer with their piezoelectric effect using VOSviewer.

In this research work, "Web of Science"–based bibliometric analysis has been performed. To collect the database, two keywords "PVDF polymer" and "piezoelectric properties" were used in this analysis. Further, a bibliometric map has been created on the basis of previously published data (Figure 3.1). In this analysis, out of 12,845 terms, 170 terms have met the condition of minimum occurrence of 15. The most suitable 100 terms have been selected for bibliographic analysis and only 46 terms have been selected for the creation of a bibliometric map. Three clusters of different colors have been (red, blue, and green) are formed per research conditions. Figure 3.2 shows the research gap analysis map for PVDF thermoplastic polymer matrix with a recyclability point of view. It shows that, much less has been reported on the recycling of feedstock filaments comprising PVDF/BTO/Gr. This literature review conducted from the Web of Science database suggests that different fields (recycling of piezoelectric polymers and its composites, 3D printing of piezoelectric sensors) are open to work where less or much less work have been reported to date.

The literature review reveals that researchers have worked on various polymer-based composites for fabrication of feedstock filaments of FDM, but hitherto little has been reported on hybrid mechanical and chemical recycling of plastics. Thus, there is a dire need to develop and manufacture the reusable polymer composites. Therefore, in this chapter the primary focus of the researcher is to determine the change in properties of developed nanocomposite after three reprocessing cycles. In this work, Gr and BTO were reinforced in the PVDF matrix via chemical-assisted mechanical blending (CAMB) for the fabrication of feedstock filaments of an existing FDM setup. Further, the prepared feedstock filaments were tested for

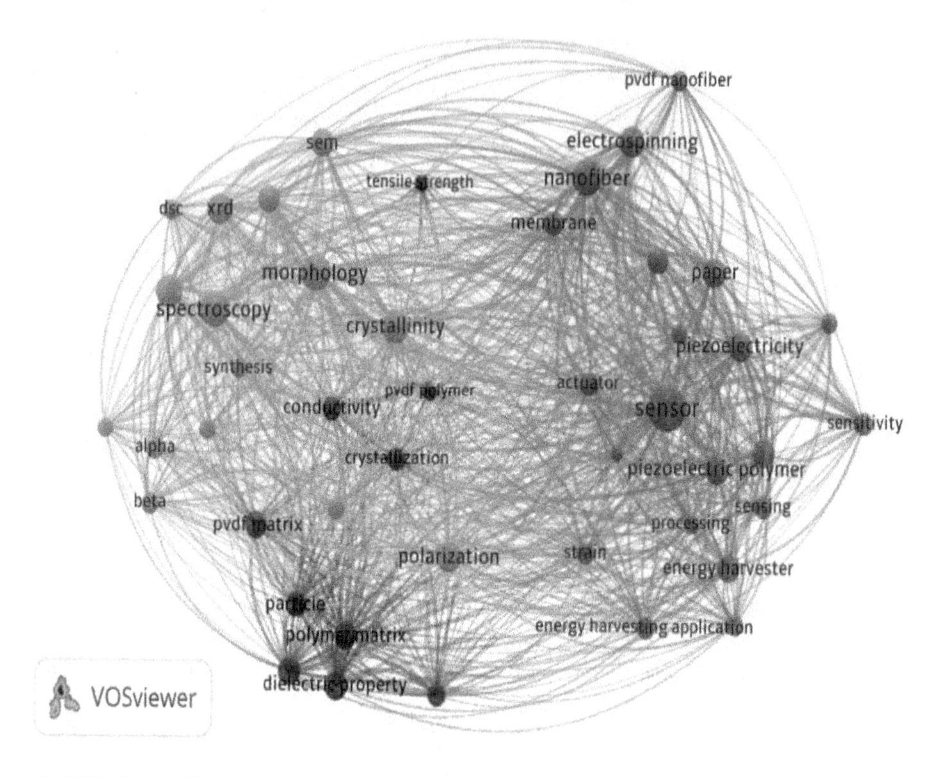

FIGURE 3.2 Bibliographic research gap analysis for PVDF polymer matrix.

mechanical properties. The fabricated feedstock filaments were recycled thrice using a twin-screw extruder to study their properties from a reusability point of view. It has been observed that for all three times prepared feedstock filaments show almost similar mechanical strengths. Thus, the developed smart material-based polymer can be reused and may work in the system of circular economy.

3.2 EXPERIMENTATION

3.2.1 Materials and Methods

Methodology used in this research work is shown in Figure 3.3.

The PVDF solef (6008/0001) was selected as a base polymer matrix. The PVDF granules used in the present study were procured from Deval Enterprises, Vadodara, India. BTO is a white ceramic powder with extraordinary piezoelectric properties. Thus, to increase the piezoelectric coefficient of base matrix, BTO with a 100 nm particle size was reinforced in it. The Gr nano-powder with a highly conductive nature was also blended in the PVDF matrix. For the preparation of composite, two different methods of blending have been adopted: (a) mechanical blending method and (b) chemical-assisted mechanical blending method. For the mechanical blending method, all three materials were mixed mechanically in a beaker with some drops of liquid adhesive.

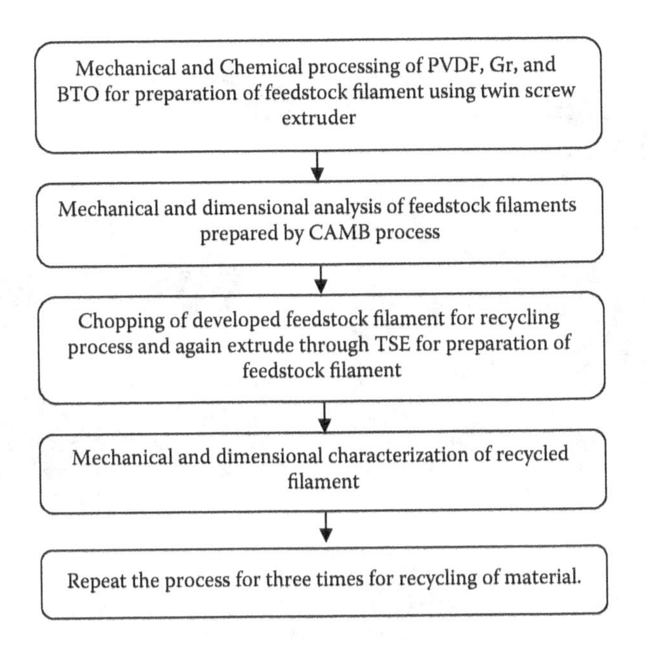

FIGURE 3.3 Methodology adopted for this research work.

In the chemical mixing method, at the first stage PVDF was dissolved in dimethyl formamide (DMF) using a hot plate magnetic stirrer. The Gr nano-powder and BTO nano-particles were ultra-sonicated in DMF at a temperature of 50°C for 30 min. Both solutions were mixed and further stirred for the next 30 min. The slurry developed through this method was poured over the glass substrate and placed inside the hot oven at 120°C for 12 hours. The setup used in chemical mixing is shown in Figure 3.4.

3.2.2 Preparation of Feedstock Filament

For preparation of feedstock filament, TSE was used. As this is the extension of previously conducted research work, therefore the same control log of experimentation (see Table 3.1) was used for the preparation of feedstock filaments of MB and CAMB composites. For each process, a total of nine sets of wires were prepared. For chemical mixing–based feedstock filaments, thin films prepared from chemical mixing were chopped and fed into the TSE (see Figure 3.5). After fabrication, feedstock filaments were subjected to mechanical testing using a universal testing machine. Further, the prepared filaments were shredded and fed into the TSE for the recycling process and again extrusion of feedstock filaments and this process was repeated a total of three times. Each time, the mechanical properties were measured to check the reusability of developed composites of smart polymer-based matrix.

3.3 RESULTS AND DISCUSSIONS

Developed feedstock filaments by using a TSE were subjected to shore hardness testing. Surface hardness was measured with a shore-D hardness tester. The

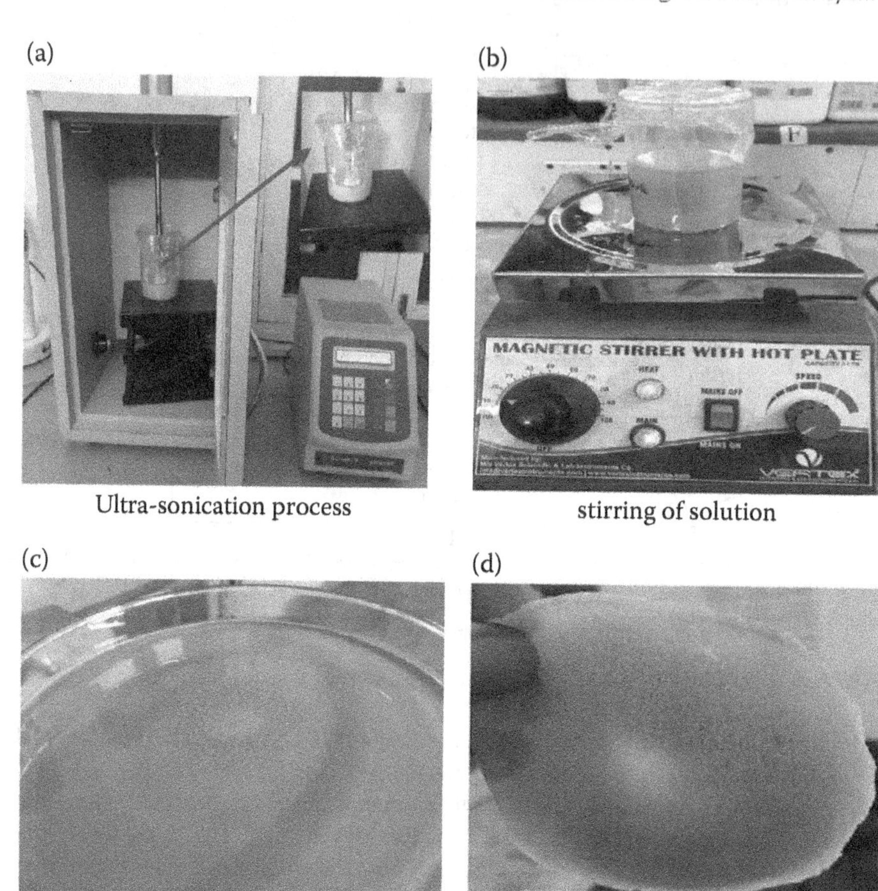

slurry poured over glass substrate for heating

Thin composite film after evaporation of DMF

FIGURE 3.4 Setup used in chemical mixing and developed composite film.

hardness of extruded filaments were found between 60 to 72 shore-D. Table 3.2 shows the outcome values of the hardness of developed filaments.

3.3.1 Mechanical Characterization

Surface hardness of the fabricated feedstock filaments were measured by using a shore-D durometer. The ANOVA table for surface hardness is shown in Table 3.3. As observed from Table 3.3, only one parameter was found significant with a P value less than 0.05, whereas the other two parameters, IA and ID, were not significant at a 95% confidence level, because the F value is not more than 20. The residual error is found at less than 4%, and represents that the model was significant.

Table 3.4 shows the rank of input variables at the condition of higher the better for observed (signal to noise) SN values of surface hardness. It has been observed

TABLE 3.1
Control log of experimentation as per Taguchi L9 orthogonal array

S. No.	Temperature	Rpm	Composition
1	180	40	A
2	180	50	B
3	180	60	C
4	190	40	B
5	190	50	C
6	190	60	A
7	200	40	C
8	200	50	A
9	200	60	B

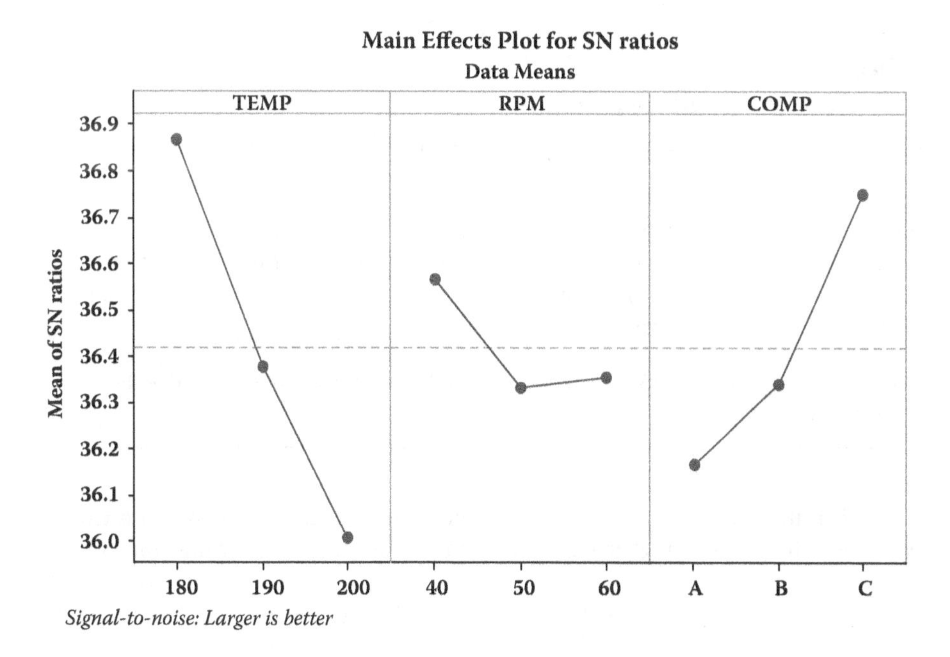

FIGURE 3.5 Graphical representation of SN ratios.

from the rank table that IS played the most prominent role in shore-D values of fabricated parts, whereas ID and IA stood 2nd and 3rd, respectively. The IA has not shown much effect on surface hardness.

Figure 3.6 shows the plot of S/N values for surface hardness. The temperature at the first level (180°C), first level of RPM (40), and third level of composition (C) have been found as best process variables settings of TSE for extrusion of feedstock filaments of PVDF+Gr+BTO composite.

TABLE 3.2
Surface hardness (shore-D) values of developed feedstock filaments

Experiment No.	Temperature °C	Rpm	Composition	Shore-D Hardness
1	180	40	A	68.22
2	180	50	B	69.14
3	180	60	C	71.94
4	190	40	B	66.42
5	190	50	C	67.12
6	190	60	A	64.19
7	200	40	C	67.44
8	200	50	A	60.72
9	200	60	B	61.48

TABLE 3.3
ANOVA table for surface hardness

Factor	DoF	Seq.SS	Adj SS	Adj MS	F	P	% Contribution
Temperature	2	1.122	1.127	0.564	23.88	0.040	61.61%
RPM	2	0.100	0.100	0.050	2.14	0.319	5.49%
Compositon	2	0.546	0.545	0.273	11.54	0.080	29.98%
Residual Error	2	0.047	0.047	0.024			2.58%
Total	8	1.821					

DoF: degree of freedom; Seq SS: sum of squares; Adj SS: adjusted sum of squares; F: Fisher value; P: probability

Further, the SN values observed have been used to calculate the optimum values of shore-D hardness at higher the better status by using the following equations.

$$\eta_{opt} = F + (F_A - F) + (F_B - F) + (F_C - F) \quad (3.1)$$

$$Z^2_{opt} = (1/10)^{\eta opt/10} \quad (3.2)$$

$$y^2_{opt} = (10)^{\eta opt/10} \quad (3.3)$$

Equations 3.2 and 3.3 are mainly used for lesser the better and higher the better respectively. So, to calculate the optimized value surface hardness (SH), the following equation has been used:

TABLE 3.4
Rank table of variable of TSE

Levels	IS	IA	ID
1	36.87	36.57	36.16
2	36.38	36.33	36.34
2	36.01	36.35	36.75
Delta	0.86	0.23	0.59
Rank	1	3	2

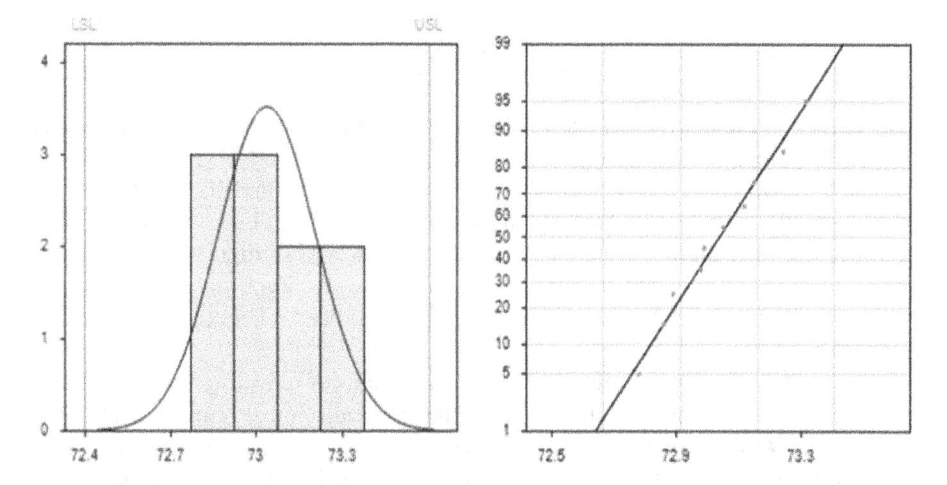

FIGURE 3.6 Histogram and normal probability plot for shore-D hardness of feedstock filament.

$$\eta_{(opt)} = F + (F_A - F) + (F_B - F) + (F_C - F)$$

$Y^2_{opt} = (10)^{\eta opt/10}$ for properties where, largest is better
F = Mean of SN for SH = 37.05
F_A = Max temperature from rank table = 36.87
F_B = Max RPM from rank table = 36.57
F_C = Max composition from rank table = 36.75
$\eta_{(opt)} = 36.41 + (36.87 - 36.41) + (36.57 - 36.41) + (36.75 - 36.41)$
$\eta_{(opt)} = 37.37$
$Y^2_{opt} = (10)^{37.92/10}$
Y_{opt} = 73.88 Shore D

The optimised value of SH is 73.88 shore-D. Further, three confirmatory experiments were conducted at the optimum conditions (i.e. temperature: 180°C, RPM 40,

TABLE 3.5
Shore-D hardness values of filaments extruded at three repetitive cycles

S. No.	R1	R2	R3
1	67.79	67.64	64.27
2	68.75	68.47	65.74
3	71.15	70.87	68.41
4	65.89	65.78	63.57
5	66.75	66.58	65.08
6	63.92	63.85	62.15
7	66.85	66.47	64.75
8	60.27	59.95	58.45
9	61.05	60.80	59.47

and composition at "C"). The experimental observed value of surface hardness is found very close to the calculated value i.e. 73.47 shore-D.

Further, to check the reusability of the developed composite, all fabricated feedstock filaments were again shredded and fed into the TSE for extrusion of filaments. All the filaments were re-extruded at all sets of settings as per the DOE. This process was repeated for three consecutive cycles. Each time after fabrication, the shore-D hardness of the filaments has been checked as shown in Table 3.5.

As seen from Table 3.5, during the first two cycles of recycling the mechanical properties of developed composites have been found almost similar to the new material; however, after the third time recycling of the material the mechanical properties start reducing. Therefore, it has been said that the developed composite of PVDF+BTO+Gr shows good thermal properties up to two repetitive thermal cycles.

3.3.2 Process Capability Analysis

Further, for process capability analysis, a total of ten samples of feedstock filaments were extruded at the obtained optimised settings. The developed feedstock filaments were subjected to shore-D hardness testing. The output results of mechanical testing of feedstock filaments in the form of shore-D values and were noted to perform process capability analysis (see Table 3.6). It has been observed from the output results of mechanical testing that less variations were found by extruding the filaments under similar processing conditions.

The process capability analysis data shown in Table 3.6 was processed with a statistical analytical tool. The calculated values of $Cp > 1$ and $Cpk > 1$, for types of specimens represented that the extruded feedstock filament was statistically under control. The outcome graphs of process capability analysis in the form of a histogram and normal probability plot for shore-D hardness of feedstock

TABLE 3.6
Shore-D hardness testing of fabricated feedstock filaments

S. No.	Shore-D Hardness
1	73.12
2	72.86
3	72.98
4	73.24
5	73.15
6	72.78
7	73.31
8	72.89
9	72.99
10	73.05

filament represents that the observed values are lying within the prescribed upper statistical limits (USLs) and lower statistical limits (LSLs). The bell-shaped curve outlined that the observed result followed the normal distribution. Further, in a normal probability plot, all the observed values of peak strength lie near the normal line, which outlined the process in control. The calculated results of process capability analysis for shore-D hardness feedstock filaments are shown in Table 3.7.

Notes:

1. *Cp and Cpk measure consistency with average performance. The "k" stands for "centralizing factor." The index takes into consideration the fact that data is maybe not centered.*
2. *For shore-D hardness, the USL and LSL were 73.6 and 72.4, respectively.*

TABLE 3.7
Results of process capability analysis for 3D-printed specimens

	Shore-D hardness
Std. Deviation	0.17219
C_p	1.16
C_{pu}	1.09
C_{pl}	1.23
C_{pk}	1.09
CR	0.86

3.4 CONCLUSIONS

This research work reports the recyclability of the in-house developed smart polymer-based composites of PVDF+BTO+Gr prepared by chemical-assisted mechanical blending method. The feedstock filaments fabricated by TSE were reprocessed three times to check their reusability for a circular economy. Following are the conclusions drawn from this research work:

- The results of ANOVA highlighted that the extruder temperature played a dominate role toward the hardness of the extruded filaments surfaces (i.e. 61.61%) followed by composition of materials that affected only 29.98%. However, screw speed was found insignificant with a P value more than 0.05.
- The results of reprocessing of developed composites show that the pvdf +bto+Gr–based composite can withstand two thermal cycles without compromising its mechanical properties. However, reprocessing of material at the third thermal cycle leads to the reduction of its mechanical properties.
- The values of Cp and Cpk obtained from the process capability analysis were found greater than 1 and clearly represents the extrusion of feedstock filaments at TSE is statistically under control process for batch production applications.

ACKNOWLEDGEMENT

The authors are extremely thankful to the Thapar Institute of Engineering and Technology, Patiala, and the Center for Manufacturing Research, Guru Nanak Dev Engineering College, Ludhiana for providing research facilities.

REFERENCES

Bengisu, M., & Ferrara, M. (2018). Materials that Move. In *Materials that Move* (pp. 5–38). Springer, Cham.

Chacón, J. M., Caminero, M. A., Núñez, P. J., García-Plaza, E., García-Moreno, I., & Reverte, J. M. (2019). Additive manufacturing of continuous fibre reinforced thermoplastic composites using fused deposition modelling: Effect of process parameters on mechanical properties. *Composites Science and Technology, 181*, 107688.

Cummings, I. T., Bax, M. E., Fuller, I. J., Wachtor, A. J., & Bernardin, J. D. (2017). A framework for additive manufacturing process monitoring & control. In *Topics in Modal Analysis & Testing, Volume 10* (pp. 137–146). Springer, Cham.

Depalma, K., Walluk, M. R., Murtaugh, A., Hilton, J., McConky, S., & Hilton, B. (2020). Assessment of 3D printing using fused deposition modeling and selective laser sintering for a circular economy. *Journal of Cleaner Production, 264*, 121567.

Deshwal, S., Kumar, A., & Chhabra, D. (2020). Exercising hybrid statistical tools GA-RSM, GA-ANN and GA-ANFIS to optimize FDM process parameters for tensile strength improvement. *CIRP Journal of Manufacturing Science and Technology, 31*, 189–199.

Dickson, A. N., Abourayana, H. M., & Dowling, D. P. (2020). 3D Printing of fibre-reinforced thermoplastic composites using fused filament fabrication—A review. *Polymers, 12*(10), 2188.

Gardan, J. (2016). Additive manufacturing technologies: state of the art and trends. *International Journal of Production Research, 54*(10), 3118–3132.

Grossule, V. (2019). Simple-Tech Solutions for Sustainable Waste Management.

Haleem, A., Javaid, M., Singh, R. P., & Suman, R. (2021). Significant roles of 4D printing using smart materials in the field of manufacturing. *Advanced Industrial and Engineering Polymer Research, 4*(4).

Jubinville, D., Esmizadeh, E., Saikrishnan, S., Tzoganakis, C., & Mekonnen, T. (2020). A comprehensive review of global production and recycling methods of polyolefin (PO) based products and their post-recycling applications. *Sustainable Materials and Technologies, 25*, e00188.

Lee, J. Y., An, J., & Chua, C. K. (2017). Fundamentals and applications of 3D printing for novel materials. *Applied Materials Today, 7*, 120–133.

Mangat, A. S., Singh, S., Gupta, M., & Sharma, R. (2018). Experimental investigations on natural fiber embedded additive manufacturing-based biodegradable structures for biomedical applications. *Rapid Prototyping Journal, 24*(7).

Martinez-Marquez, D., Mirnajafizadeh, A., Carty, C. P., & Stewart, R. A. (2018). Application of quality by design for 3D printed bone prostheses and scaffolds. *PLoS One, 13*(4), e0195291.

Maurya, D., Peddigari, M., Kang, M. G., Geng, L. D., Sharpes, N., Annapureddy, V., Palneedi, H., Sriramdas, R., Yan, Y., Song, H. C., & Wang, Y. U. (2018). Lead-free piezoelectric materials and composites for high power density energy harvesting. *Journal of Materials Research, 33*(16), 2235–2263.

Melly, S. K., Liu, L., Liu, Y., & Leng, J. (2020). On 4D printing as a revolutionary fabrication technique for smart structures. *Smart Materials and Structures, 29*(8), 083001.

Moore, C. J. (2008). Synthetic polymers in the marine environment: a rapidly increasing, long-term threat. *Environmental Research, 108*(2), 131–139.

Ngo, T. D., Kashani, A., Imbalzano, G., Nguyen, K. T., & Hui, D. (2018). Additive manufacturing (3D printing): A review of materials, methods, applications and challenges. *Composites Part B: Engineering, 143*, 172–196.

Park, S., & Fu, K. K. (2021). Polymer-based filament feedstock for additive manufacturing. *Composites Science and Technology*, 213, 108876.

Ramya, A., & Vanapalli, S. L. (2016). 3D printing technologies in various applications. *International Journal of Mechanical Engineering and Technology, 7*(3), 396–409.

Safaei, M., Sodano, H. A., & Anton, S. R. (2019). A review of energy harvesting using piezoelectric materials: state-of-the-art a decade later (2008–2018). *Smart Materials and Structures, 28*(11), 113001.

Sharma, R., Singh, R., Penna, R., & Fraternali, F. (2018). Investigations for mechanical properties of Hap, PVC and PP based 3D porous structures obtained through biocompatible FDM filaments. *Composites Part B: Engineering, 132*, 237–243.

Sharma, R., Singh, R., & Batish, A. (2019). Study on barium titanate and graphene reinforced PVDF matrix for 4D applications. *Journal of Thermoplastic Composite Materials*, 0892705719865004.

Sharma, R., Singh, R., & Batish, A. (2020). On effect of chemical-assisted mechanical blending of barium titanate and graphene in PVDF for 3D printing applications. *Journal of Thermoplastic Composite Materials*, 0892705720945377.

Sharma, R., Singh, R., & Batish, A. (2020). On mechanical and surface properties of electro-active polymer matrix-based 3D printed functionally graded prototypes. *Journal of Thermoplastic Composite Materials*, 0892705720907677.

Sharma, R., Singh, R., Batish, A., & Ranjan, N. (2021). On synergistic effect of BTO and graphene reinforcement in polyvinyl diene fluoride matrix for four dimensional applications. *Proceedings of the Institution of Mechanical Engineers, Part C: Journal of Mechanical Engineering Science*, 09544062211015763.

Sharma, R., Singh, R., & Batish, A. (2020). Investigations for Barium Titanate and Graphene Reinforced PVDF Matrix for 4D Applications.

Sharma, R., Singh, R., & Batish, A. (2020). On multi response optimization and process capability analysis for surface properties of 3D printed functional prototypes of PVC reinforced with PP and HAp. *Materials Today: Proceedings*, *28*, 1115–1122.

Sharma, R., Singh, R., & Batish, A. (2021). On flexural and pull out properties of smart polymer based 3D printed functional prototypes. *Sādhanā*, *46*(3), 1–18.

Shanmugam, V., Das, O., Neisiany, R. E., Babu, K., Singh, S., Hedenqvist, M. S., Berto, F., & Ramakrishna, S. (2020). Polymer recycling in additive manufacturing: An opportunity for the circular economy. *Materials Circular Economy*, *2*(1), 1–11.

Shekdar, A. V. (2009). Sustainable solid waste management: An integrated approach for Asian countries. *Waste Management*, *29*(4), 1438–1448.

Singh, R., Sharma, R., & Davim, J. P. (2018). Mechanical properties of bio compatible functional prototypes for joining applications in clinical dentistry. *International Journal of Production Research*, *56*(24), 7330–7340.

Singh, S., Singh, G., Prakash, C., & Ramakrishna, S. (2020). Current status and future directions of fused filament fabrication. *Journal of Manufacturing Processes*, *55*, 288–306.

Singh, R., Kumar, R., & Singh, I. (2021). Investigations on 3D printed thermosetting and ceramic-reinforced recycled thermoplastic-based functional prototypes. *Journal of Thermoplastic Composite Materials*, *34*(8), 1103–1122.

Verma, R., Vinoda, K. S., Papireddy, M., & Gowda, A. N. S. (2016). Toxic pollutants from plastic waste-a review. *Procedia Environmental Sciences*, *35*, 701–708.

Zafar, M. Q., & Zhao, H. (2020). 4D printing: future insight in additive manufacturing. *Metals and Materials International*, *26*(5), 564–585.

4 Primary and Secondary Melt Processing for Plastics

Kamaljit Singh Boparai, Abhishek Kumar, and Rupinder Singh

4.1 INTRODUCTION

Additive manufacturing (AM) is considered a vital ingredient in making Industry 4.0 a successful model. It has been reported that Industrial Revolution 4.0 has fundamentally transformed the production approach in which AM revolutionized both designing and manufacturing processes in industrial and domestic applications due to its distinctive competencies and capability to 3D design complex structures with promising high-grade mechanical, thermal, and physical properties (Gibson et al., 2014, Gao et al., 2015). Presently, the AM process has evolved from a prototype building technique for product development to functional part fabrication in diverse applications like automotive, aerospace, food technology, biomedical science, industrial and special machinery, and robotics (Gardan, 2016; Garg & Bhattacharya, 2017; Boparai et al., 2021). As per ASTM F2792, seven domains of AM are described as VAT photo-polymerization, powder bed fusion (PBF), binder jetting, material jetting, sheet lamination, material extrusion, and direct energy deposition. The FDM belongs to the material extrusion classification of the AM, which uses consumable feedstock filament material as raw material based on polymeric material. The feedstock filament by the action of roller mechanism is passed through the liquefier hot head in a molten state to a predetermined path to fabricate a single part. David et al. (2014) investigated the part-building process by varying process parameters: layer thickness and road width on surface metrology, production times consumed and influence on mechanical properties for a FDM system and additionally developed a novel FDM system that enables the user to deposit separate multiple materials within a selected region. The part built with this system has shown less production time with improved surface and mechanical characteristics. Onwubolu and Rayegani (2014) studied the FDM process for acrylonitrile butadiene styrene (ABS) parts to observe the influence on mechanical properties and concluded that tensile stress can be increased by printing part with low thickness. Ziemian et al. (2015) outlined that FDM-printed ABS parts have anisotropic properties and summarized that parts printed with a 0° raster angle have better tensile properties. Torrado Perez et al. (2014) investigated the influence on

DOI: 10.1201/9781003184164-4

mechanical properties of ABS reinforced with TiO_2 and observed that with reinforcement the ultimate tensile strength was increased in comparison to virgin ABS.

It has been reported that properties of an AM-based printed part is mainly dependent on the mechanical and structural properties of feedstock filament material. Thermoplastic material can be recycled and changed to a new shape without any significant change in structural property and provides advantages such as easy processability, low cost, and low melting temperature but lacks physical strength for functional applications (Shofner et al., 2003, Nikzad et al., 2011). A thermosetting polymer matrix provides promising properties such as high strength, lightweight, processability, and flexibility, but it has some disadvantages like poor mechanical properties and low printing speed (Boparai & Singh, 2019a,b). Thermoplastic-based feedstock filament reinforced with additives provides high fatigue strength and improved mechanical and physical characteristics (Wang et al., 2017). Due to rapid growth and advancement of AM in diverse applications, the need of developing feedstock filament with high performance and properties has shown an exceptional rise (Boparai et al., 2016a,b). The necessity of developing feedstock filament with enhanced mechanical properties for engineering applications in widespread fields cannot be overlooked. In the past two decades, researchers and academicians have been working to find alternatives to replace metals from structural applications, encouraging lightweight structures, low cost, machineability, and feasibility. The development in area of FDM materials with high processability, good mechanical properties, and promising characteristics can increase the utilization of the FDM part-building process more efficiently and rapidly in functional and nonfunctional applications (Roberson et al., 2015). Biodegradable polymers are now extensively used in emerging areas of tissue engineering to observe the potential of polymer scaffolds to aid the cell and tissue growth during genesis of artificial organs (Taboas et al., 2003). Kuang et al. (2018) developed a two-stage curing process-based hybrid approach to fabricate thermoset feedstock filament-based parts at a high print speed because with a conventional method it was impossible to build parts with thermosets.

The presented work provides comparative analysis between the experimental and computational approach, finite element simulation and machine learning, to investigate the tensile properties of alternative feedstock filament thermoplastic material as binder material with reinforced metal additives as filler material. The developed feedstock filament may be utilized in lieu of standard commercial feedstock filament materials available in the market.

4.2 EXPERIMENTATION

4.2.1 Material Selection

In this case study, three materials were selected for the fabrication of alternative feedstock filament, Nylon6 as binder material and Al and Al_2O_3 as filler material. Nylon6 (E-35 grade) was procured from Gujrat State Fertilizer Limited, India with a mass density of 1.14 g/cm³. It has numerous applications such as parachutes,

TABLE 4.1
Properties of Nylon6 (supplier's data)

S No.	Properties	Test Method	Units
1	Tensile strength	ASTM D638	700 ± 50 kg/cm^2
2	Elongation at yield	do	5 mm
3	Elongation at break	do	250 ± 20 mm
4	Flexural strength	ASTM D790	875 ± 50 kg/cm^2
5	Flexural modulus	do	21–23 kg/cm^2
6	Izod impact strength	do	3–4.5 kg.cm/cm of notch
7	Rockwell hardness	ASTM D785	120 ± 5 RHR
8	Heat deflection	ASTM D648	60 ± 2 °C

Source: Gujrat State Fertilizer Limited.

industrial cords, conveyor belts, and swim wear, etc. The properties of Nylon6 are shown in Table 4.1.

Al is a lightweight metal with a low-density 2.7 g/cm^3 procured from Thomas Bakers, India, was selected as filler material. Al is selected because of its exceptional properties such as good mechanical and physical strength, good heat conductivity, and high ductility. It finds wide application in aerospace, automotive, and transportation due to its lightweight nature. In addition to Al, Al_2O_3 having a density of 3.95 g/cm^3 was also procured from Thomas Bakers, India, and selected as filler material for the preparation of composition. The important properties of both filler material have been shown in Table 4.2.

4.2.2 Material Processing

The binder material (Nylon6) and filler materials (Al and Al_2O_3) were separately heated and maintained at absolute zero pressure at 50°C for 10 hours using a

TABLE 4.2
Properties of filler materials (supplier's data)

S. No.	Property	Al	Al_2O_3
1	Melting point	660.32°C	2075
2	Molar heat capacity	24.200 J·mol^{-1}·K^{-1}	–
3	Density	2.7 g/cm^3	3.95–4.1 g/cm^3
4	Electrical resistivity	At 20°C 28.2 nΩ-m	–
5	Thermal conductivity	237 W-m^{-1}k^{-1}	30 W-m^{-1}k^{-1}
6	Thermal expansion	At 25°C 23.1 μm-m^{-1}k^{-1}	–
7	Young's modulus	70 GPa	–

TABLE 4.3
Weight proportion of compositions

Composition	Nylon6	Al	Al_2O_3
A	60	26	14
B	60	28	12
C	60	30	10

vacuum oven. It was important and crucial to remove the moisture content from both binder and filler material so that during processing of material for the feedstock filament fabrication in screw extruder no oxidation takes place in the material. The varying weight proportion (Table 4.3) of binder and filler material was prepared by using a mechanical mixer (at 200 rpm for 2 hours). Due to good binding characteristics of Nylon6 and self-lubricating property of Al, no surfactants or plasticizer were added in the mixture.

4.2.3 Sample Preparation

Design of experiment (DOE) was used to create a log of experiments (Table 4.4).

The 3-level and 4-factor-based Taguchi L9 orthogonal array (OA) was prepared, as shown in Table 4.5.

A single-screw extruder has been used to fabricate the feedstock filament. The prepared material composition was fed in the hopper and processed through varying process parameters. The characteristics of the feedstock filament is mainly dependent on the mechanical and physical properties of material and processing conditions. Further, the fabricated feedstock filament is inspected for dimensions and observed that the diameter of the filament is in accordance with a standard filament diameter of 1.75 mm. The prepared feedstock filament specimen was spooled and stored in a dry environment. Figure 4.1 shows the prepared feedstock filament.

4.2.4 Uniaxial Tensile Testing

The compositions of prepared feedstock filament were tested according to the ASTM-638 standard on the universal testing machine (UTM) for the purpose of

TABLE 4.4
Process parameters

Level	Material composition	Mean Barrel Temp.	Die Temp.	Screw Speed
1	A	160	185	25
2	B	170	195	30
3	C	180	205	35

TABLE 4.5
L9 OA experiment design

Experiment runs	Composition	Mean Barrel Temperature °C	Die Temperature °C	Screw Speed (rpm)
1	A	160	185	25
2	A	170	195	30
3	A	180	205	35
4	B	160	195	35
5	B	170	205	25
6	B	180	185	30
7	C	160	205	30
8	C	170	185	35
9	C	180	195	25

FIGURE 4.1 FDM feedstock filament spool.

investigating the tensile properties under structural loading conditions. Polymeric material depicts the ductile and brittle behavior as metallic components. In addition to this, they also show a similar behavior when under loading conditions in a highly elastic manner (Berns, 1991). The tensile properties of polymeric materials were also evaluated in a similar manner as metallic materials using a universal tensile testing machine. Gere and Timoshenko (1990) summarized that when polymer parts under loading conditions fail due to induction of yield stress and yield strain in the material leads to a reduction in ability of the material to carry the load. The specimen part should be inspected for any defects before a uniaxial tensile test because when the tensile starts begin with application of load the buildup stress should be uniform throughout the gauge section of the specimen. If the specimen has some defects due to molecular structure, surface defect, or contaminate the stress, distribution will not be uniform and stress distribution will not be on the other side of

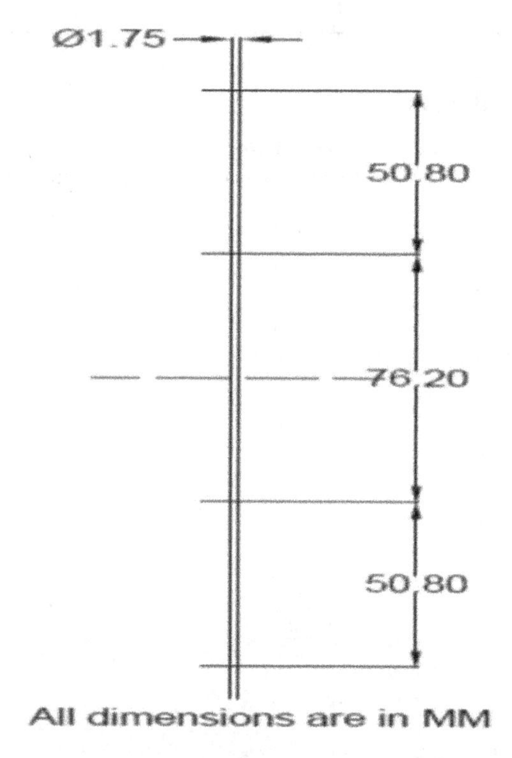

FIGURE 4.2 Specimen dimension.

the surface with the defect (Courtney, 1990, Joun et al., 2007). Figure 4.2 shows the dimension of the prepared filament specimen standard for the tensile testing and Figure 4.3 shows the tensile testing machine.

4.2.5 Finite Element Analysis

It is crucial to develop finite element models for reducing the cost of conducting experiments, reducing material wastage to investigate the elastoplastic deformation behavior (Schoinochoritis et al., 2017). In the presented study, finite element analysis (FEA) was carried out using the ABAQUS simulation software package. FEA can also be used to optimize the structure and process parameters. The objective of FEA was to visualize the behavior of material under a loading condition and develop a correlation approach between the experimental uniaxial tensile testing and finite element modeling. FEA consists of three main steps: pre-processing includes geometry, material assignment, and boundary conditions; processing includes calculations of the results; post-processing includes the validation and analysis of the results. In the pre-processing stage, first a tensile testing specimen geometry is modelled with the dimensions shown in Figure 4.4. Material properties were assigned from the observed experimental data to the modelled geometry, such as mass density, true stress-strain, and plastic strain. The mass density of composition is given by

FIGURE 4.3 Universal tensile testing machine.

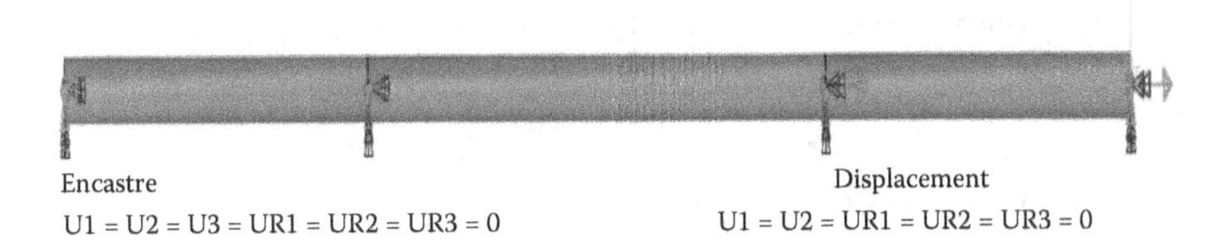

FIGURE 4.4 Boundary condition.

Density of composition: $\rho c = 1 / \left\{ \left(\frac{Wn}{\rho n} \right) + \left(\frac{WAl}{\rho Al} \right) + \left(\frac{Wa}{\rho a} \right) \right\}$ [Shenoy et al., 1984]

where

- ρc = density of composition; ρn = density of Nylon6; ρAl = density of aluminium; ρa = density of Al_2O_3

- Wn = weight proportions of Nylon6; WAl = weight proportions of aluminium; Wa = weight proportions of Al_2O_3

$$\textit{Truestress} = (\textit{engineeringstress}) * (1 + \textit{engineeringstrain})$$

$$\textit{Truestrain} = \ln(1 + \textit{engineeringstrain})$$

$$\textit{Plasticstrain} = \textit{Truestrain} - \frac{\textit{Truestress}}{\textit{youngmodulus}}$$

The ductile damage initiation plasticity model (ABAQUS version 6.6 documentation) is used, which requires input parameters for studying the material fracture, such as fracture strain, stress triaxiality, strain rate, and displacement at failure. The FEA simulation on the specimen was carried out using a dynamic explicit method. The boundary conditions (BC 1) were assigned with the command "Encastre" U1 = U2 = U3 = UR1 = UR2 = UR3 = 0 to replicate the fixed fixture in tensile testing. The "Encastre" command fixes and holds the one end and all 6 degrees of freedom is set to zero to stop the displacement and angular rotation. The boundary condition (BC 2) was also assigned the command "Displacement" U1 = U2 = UR1 = UR2 = UR3 = 0 and U3 = "Displacement value" to replicate the movable fixture in tensile testing (shown in Figure 4.4). Further, the mesh element size selected was 0.25.

Garg and Bhattacharya (2017) presented the effect of layer thickness and raster angle by developing realistic finite element using ABAQUS and validated the results experimentally. Similarly (Hussin et al., 2020), using the elasto-plastic method simulated the thermoplastic material with partial infill patterns using ABAQUS. The observations in the present case are in line with these investigators.

4.3 RESULTS AND DISCUSSION

4.3.1 Uniaxial Tensile Testing

To observe the mechanical properties of prepared feedstock filament, the uniaxial tensile testing is carried out using a universal testing machine. The tensile testing was done on all the prepared samples. It was observed that tensile properties of Nylon6 decrease with reinforcement of Al and Al_2O_3 in the matrix. The variation in weight proportion of binder and filler materials should maintain at a level limit. The prepared feedstock filament should have adequate mechanical strength so that it can be processed without getting broken or buckling inside the extruder head. It was observed that the result of the tensile test varies between the limit range of (20.80–21.75 MPa). In addition to this, if the filler material in the binder material is not maintained up to a limit, then it will reduce the melt flow index value, which can lead to choking of feedstock filament inside the liquefier head or the print quality. Figure 4.5 shows a stress-strain curve observed from uniaxial testing. Table 4.6 shows the experimental tensile results.

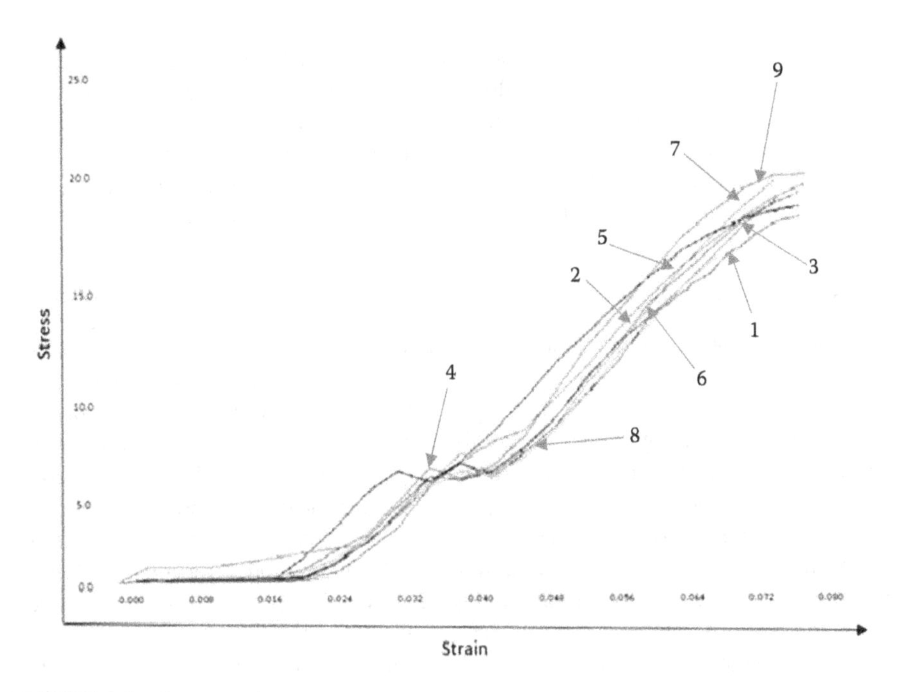

FIGURE 4.5 Stress-strain curve.

TABLE 4.6
Tensile testing experimental outputs

Experiment No.	Tensile strength (MPa)			Mean
	Trial 1	Trial 2	Trial R3	
1	20.95	20.80	20.97	20.9067
2	21.10	21.11	21.15	21.1200
3	21.40	21.52	21.30	21.4067
4	21.53	21.48	21.50	21.5033
5	21.43	21.40	21.55	21.4600
6	21.48	21.46	21.50	21.4800
7	21.60	21.40	21.75	21.5833
8	21.25	21.28	21.19	21.2400
9	21.65	21.69	21.60	21.6467

4.3.2 Finite Element Analysis

The purpose of finite element analysis is to develop a computational approach to study the elastoplastic of material. Finite element analysis replicates the behaviour of material under loading conditions with realistic conditions. To validate the FEA study, the percentage of deviation between the experimental test and simulation test must be below 10% (Zhang et al., 2007). The ductile damage criterion was applied to replicate the homogenous plastic flow of material under the loading condition. It has been visualised that a fracture in all the specimens was at the displacement range of 0.02–0.28 mm from the beginning of the gauge section. Figure 4.6(a-i) depicts the simulation behaviour of the specimen and Table 4.7 shows the deviation between the experimental results and finite element simulation result.

4.3.3 Machine Learning

Machine learning is a field of artificial intelligence in which computational technology in conjunction with statistical tools are used to compute, optimize, and predict an ideal process (Li et al., 2019) developed with a machine learning–based approach by processing data using the RF method before training the ML model to predict the surface roughness of parts printed with the FDM process. Ogorodnyk et al. (2018) used an artificial neural network (ANN)–based machine learning approach to separate the defected parts from the ideal parts processed through injection moulding. Donegan et al. (2020) developed a machine learning model for metal additive manufacturing using unsupervised learning to establish a relation between structures and process of different components. Baturynska et al. (2018) used MLP- and CNN-based machine learning to optimise and predict the dimensional accuracy in the SLS process. In the presented work, a simple multivariate regression-based prediction model is developed to predict the tensile stress of alternative feedstock filament prepared using a Nylon6-Al-Al_2O_3 composition. The presented machine learning model is based on supervised learning, in which data is trained with the dependent variable and independent variable i.e. material composition, mean barrel temperature, die temperature, screw speed, tensile stress. In the study, tensile stress is provided as an independent variable and the process variable as a dependent variable. Figure 4.7 shows the developed machine learning model for the prediction of tensile stress. The developed model uses Python libraries such as panda for data frame analysis, NumPy, and SkLearn for the mathematical computation and statistics.

The multiple regression equation formulated for the study:

$$Tensile\ \ stress = m1 * material\ \ Composition + m2 * mean\ barrel\ \ temperature \\ + m3 * die\ \ temperature + m4 * screw\ \ speed + b$$

where

m = coefficient
b = intercept

The developed machine learning model was trained such that by calling the process variable value i.e. material composition, mean barrel temperature, die temperature, screw speed in the prediction bar we can predict the approximate value of the process. Figure 4.7 shows the predicted model for calculating the values of tensile strength by varying the proportion of Nylon6-Al-Al_2O_3–based composition for the production of alternative feedstock filament for FDM (Figure 4.8).

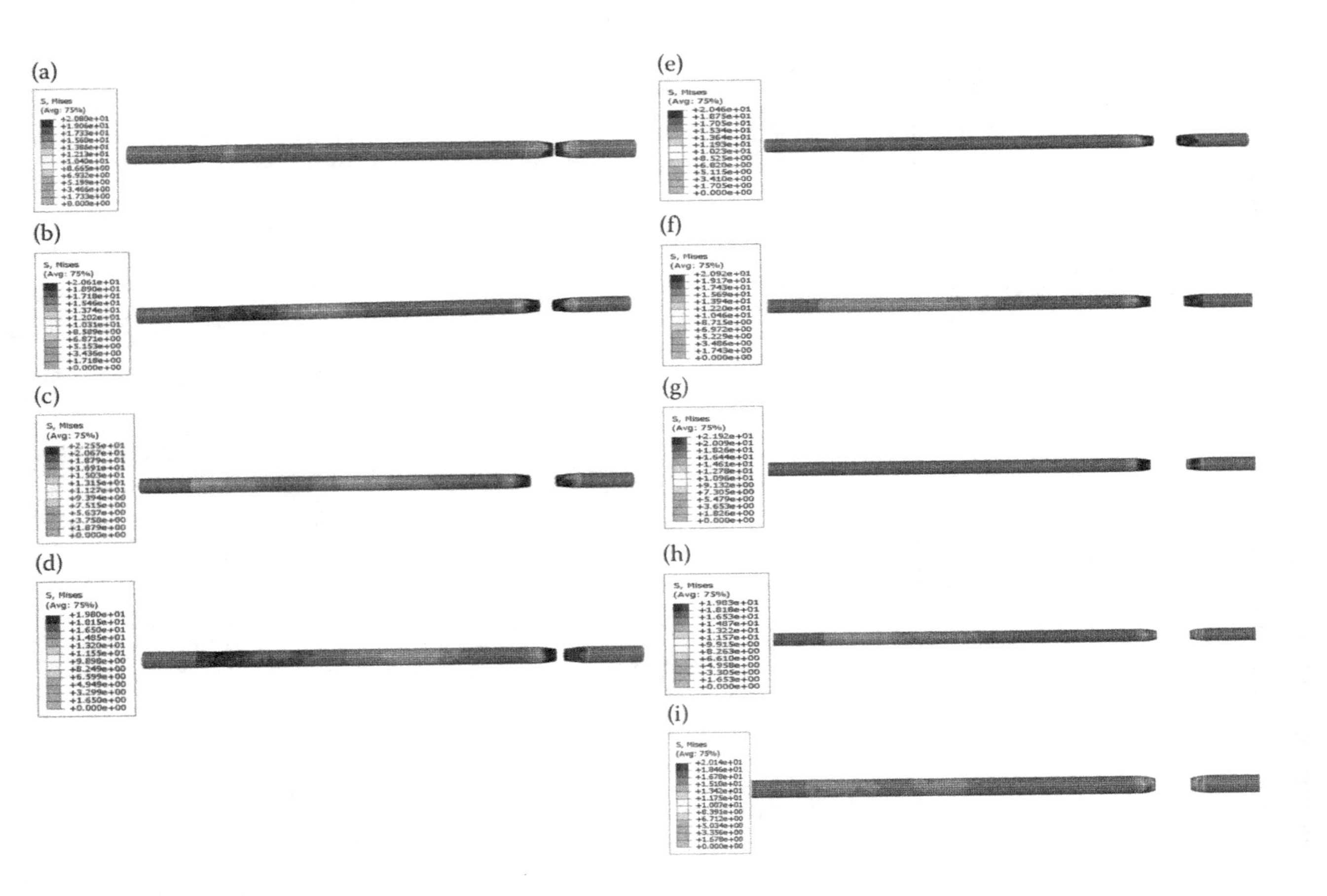

FIGURE 4.6 Finite element analysis (a-i).

TABLE 4.7
Experimental and FEA deviation

Experiment run	Experimental tensile stress (MPa)	FEA tensile stress (MPa)	Deviation
1	20.90	21.80	−0.9
2	21.12	21.69	−0.57
3	21.40	20.97	0.43
4	21.50	20.18	1.32
5	21.46	20.91	0.55
6	21.48	21.03	0.45
7	21.58	22.76	−1.18
8	21.24	20.39	0.85
9	21.64	21.11	0.53

```
In [11]: import pandas as pd
         import numpy as np
         import sklearn as sk
         from sklearn import linear_model

In [12]: df= pd.read_csv('book1.csv')
         df

Out[12]:
```

	composition	screw_speed	barrel_temperature	die_temperature	tensile_stress
0	0	160	185	25	20.9067
1	0	170	195	30	21.1200
2	0	180	205	35	21.4067
3	1	160	195	35	21.5033
4	1	170	205	25	21.4600
5	1	180	185	30	21.4800
6	2	160	205	30	21.5833
7	2	170	185	35	21.2400
8	2	180	195	25	21.6467

```
In [13]: #Model to train the machine to predict the outputs
         reg = linear_model.LinearRegression()
         #reg.fit(input_value,output_values)
         reg.fit(df[['composition','screw_speed','barrel_temperature','die_temperature']],df['tensile_stress'])

Out[13]: LinearRegression()
```

FIGURE 4.7 Machine learning model.

```
In [14]: #output
         predicted_value = reg.predict([["composition value","screw_speed value","barrel_temperature value","die_temperature value"]])

         print(predicted_value)
```

FIGURE 4.8 Prediction model.

CONCLUSIONS

The following conclusions may be drawn for the present research work:

i. The Nylon6-Al-Al_2O_3–based composition can be recommended for the production of alternative feedstock filament in place of ABS filament.
ii. The process variable such as material composition, mean barrel temperature, die temperature, and screw speed have a strong influence on the mechanical properties of feedstock filament.
iii. The FEA study illustrates the percentage of deviation (which is less than 10%) between the experimental test and simulation test.
iv. The Python-based machine learning approach developed a predicted model for calculating the values of tensile strength by varying the proportion of the Nylon6-Al-Al_2O_3–based composition.

REFERENCES

ABAQUS, V. (2006). 6.6, User Documentation, Dassault Systems.

ASTM F2792 (2012). *ASTM F2792 – 12a: Standard Terminology for Additive Manufacturing Technologies*, ASTM International, West Conshohocken, PA. Available online at: www.astm.org

Baturynska, I., Semeniuta, O., & Wang, K. (2018). September. Application of machine learning methods to improve dimensional accuracy in additive manufacturing. In *International Workshop of Advanced Manufacturing and Automation* (pp. 245–252). Springer, Singapore.

Boparai, K. S., & Singh, R. (2019a). Recyclability of Packaging Materials for Domestic Applications.

Boparai, K. S., & Singh, R. (2019b). 3D printed functional prototypes for electrochemical energy storage. *International Journal of Materials Engineering Innovation*, *10*(2), 152–164.

Boparai, K. S., Kumar, A., Kumar, A., Aman, A., & Singh, S. (2021). Nanomaterial in additive manufacturing for energy storage applications. In *Handbook of Polymer Nanocomposites for Industrial Applications* (pp. 529–543). Elsevier, New Jersey, USA.

Boparai, K. S., Singh, R., & Singh, H. (2016a). Development of rapid tooling using fused deposition modeling: a review. *Rapid Prototyping Journal*, 22 (2).

Boparai, K. S., Singh, R., Fabbrocino, F., & Fraternali, F. (2016b). Thermal characterization of recycled polymer for additive manufacturing applications. *Composites Part B: Engineering*, *106*, 42–47.

David E., Jorge, A. R., Francisco, M., & Ryan, W. (2014). Multi-material, multi-technology FDM: exploring build process variations, *Rapid Prototyping Journal*, *20*(*3*), 236–244.

Donegan, S. P., Schwalbach, E. J., & Groeber, M. A. (2020). Zoning additive manufacturing process histories using unsupervised machine learning. *Materials Characterization*, *161*, 110123.

Gao, W., Zhang, Y., Ramanujan, D., Ramani, K., Chen, Y., Williams, C. B., Wang, C. C., Shin, Y. C., Zhang, S., & Zavattieri, P. D. (2015). The status, challenges, and future of additive manufacturing in engineering. *Computer-Aided Design*, *69*, 65–89.

Gardan, J. (2016). Additive manufacturing technologies: state of the art and trends. *International Journal of Production Research*, *54*(10), 3118–3132.

Garg, A., & Bhattacharya, A. (2017). An insight to the failure of FDM parts under tensile loading: finite element analysis and experimental study. *International Journal of Mechanical Sciences*, *120*, 225–236.

Gibson, I., Rosen, D., Stucker, B., & Khorasani, M. (2014). *Additive manufacturing technologies* (*Vol. 17*, p. 195). New York: Springer.

Hussin, M. S., Hamat, S., Ali, S. A. S., Fozi, M. A. A., Rahim, Y. A., Dawi, M. S. I. M., & Darsin, M. (2020). October. Experimental and finite element modeling of partial infill patterns for thermoplastic polymer extrusion 3D printed material using elasto-plastic method. In *AIP Conference Proceedings* (*Vol. 2278*, 1, p. 020011). AIP Publishing LLC.

James M. Gere, & Stephen P. Timoshenko (1990). *Mechanics of Materials*. 3rd Edition, New York, PWSKENT Publishing Co.

Joun, M., Choi, I., Eom, J., & Lee, M. (2007). Finite element analysis of tensile testing with emphasis on necking. *Computational Materials Science*, *41*(1), 63–69.

Kuang, X., Shi, Q., Zhou, Y., Zhao, Z., Wang, T., & Qi, H. J. (2018). Dissolution of epoxy thermosets via mild alcoholysis: the mechanism and kinetics study. *RSC advances*, *8*(3), 1493–1502.

Li, Z., Zhang, Z., Shi, J., & Wu, D. (2019). Prediction of surface roughness in extrusion-based additive manufacturing with machine learning. *Robotics and Computer-Integrated Manufacturing*, *57*, 488–495.

Michael L. Berns (1991) *Plastics Engineering Handbook of the Plastics Industry*, 5th Edition, New York, Van Norstrand Reinhold.

Nikzad, M., Masood, S. H., & Sbarski, I. (2011). Thermo-mechanical properties of a highly filled polymeric composites for fused deposition modeling. *Materials & Design*, *32*(6), 3448–3456.

Ogorodnyk, O., Lyngstad, O. V., Larsen, M., Wang, K., & Martinsen, K. (2018), September. Application of machine learning methods for prediction of parts quality in thermoplastics injection molding. In *International workshop of advanced manufacturing and automation* (pp. 237–244). Springer, Singapore.

Onwubolu, G. C., & Rayegani, F. (2014). Characterization and optimization of mechanical properties of ABS parts manufactured by the fused deposition modelling process. *International Journal of Manufacturing Engineering, 2014.*

Roberson, D., Shemelya, C. M., MacDonald, E., & Wicker, R. (2015). Expanding the applicability of FDM-type technologies through materials development, *Rapid Prototyping Journal*, *21*(2), 137–143.

Schoinochoritis, B. , Chantzis, D. , & Salonitis, K. (2017). Simulation of metallic powder bed additive manufacturing processes with the finite element method: A critical review. *Proceedings of the Institution of Mechanical Engineers, Part B: Journal of Engineering Manufacture*, 231(1), 96–117.

Shenoy, A. V., Saini, D. R., & Nadkarni, V. M. (1984). Melt rheology of polymerblends from melt flow index. *International Journal of Polymeric Materials*, *10*(3) 213–235.

Shofner, M. L., Lozano, K., Rodríguez-Macías, F. J., & Barrera, E. V. (2003). Nanofiber-reinforced polymers prepared by fused deposition modeling. *Journal of Applied Polymer Science*, *89*(11), 3081–3090.

Standard, B. and ISO, B. (1996). Plastics—Determination of tensile properties—. Part, 1, pp. 527–521.

Taboas, J. M., Maddox, R. D., Krebsbach, P. H., & Hollister, S. J. (2003). Indirect solid free form fabrication of local and global porous, biomimetic and composite 3D polymer-ceramic scaffolds. *Biomaterials*, *24*(1), 181–194.

Thomas H. Courtney (1990). *Mechanical Behavior of Materials*, New York, McGraw-Hill.

Torrado Perez, A. R., Roberson, D. A., & Wicker, R. B. (2014). Fracture Surface Analysis of 3D-Printed Tensile Specimens of Novel ABS-Based Materials, *Journal of Failure Analysis and Prevention*, *14*(3), 343–353.

Wang, X., Jiang, M., Zhou, Z., Gou, J., & Hui, D. (2017). 3D printing of polymer matrix composites: A review and prospective. *Composites Part B: Engineering*, *110*, 442–458.

Zhang, Z., Zhang, W., Zhai, Z. J., & Chen, Q. Y. (2007). Evaluation of various turbulence models in predicting airflow and turbulence in enclosed environments by CFD: Part 2—Comparison with experimental data from literature. *Hvac&R Research*, *13*(6), 871–886.

Ziemian, S., Okwara, M., & Ziemian, C. W. (2015). Tensile and fatigue behavior of layered acrylonitrile butadiene styrene. *Rapid Prototyping Journal*, *21*(3).

5 Fused Deposition Modeling as a Secondary Recycling Process for the Preparation of Sustainable Structures

Jaspreet Singh, Kapil Chawla, and Rupinder Singh

5.1 INTRODUCTION

The enormous applications of plastics in daily life make them an essential and vital part of human life. Their unique features, such as low density, low weight, high strength, low cost, and easy processing, makes them indispensable materials (Mwanza & Mbohwa, 2017). But their repeated, long-chain, and stable polymeric structure does not allow them to degrade easily in natural conditions and they remain in the same form for hundreds of years. Landfilling is commonly employed for polymeric waste (PW) in many countries. A rapid increase of PW induces a number of critical socio-environmental complications such as reduction in geographic space due to landfills, terrestrial and marine pollution, and adverse effect on biological life and freshwater resources, etc (Kozderka et al., 2016, Al-Mulla and Gupta 2018, Jambeck et al., 2018, Silvarrey & Phan, 2016, Geyer et al., 2017, Cole et al., 2011). These problems mainly arise due to low cost and unmatchable applications of polymeric materials, especially in the packaging industry where they are generally used once or for a short service span. In order to generate virgin plastics, 4% of the world's oil production is consumed each year (Singh et al., 2017). It has been reported that 10% of the PW comes from municipal waste, which ultimately increases the demand of open space for landfilling (Milios and Eriksen et al., 2018). In 2018, the plastic production reached the value of 359 million tons globally that increased approximately 20 times in the last 50 years (Milios et al., 2018). The variation in consumption and production of plastic products that in turn depends upon demand and supply also influences the generation of PW.

The worldwide existing polymers are broadly classified in three categories: elastomers, thermoplastics (materials that can be recycled up to certain extent), and thermosetting (that cannot be recycled) (da Silva & Wiebeck, 2020). Among all these materials, thermoplastics address the significant gathering and replacing a

DOI: 10.1201/9781003184164-5

large portion of the materials in different applications ranging from packaging, automotive, medicine, construction, and agriculture, as well as in the electronic industry (Milios et al., 2018). Thermosettings are monomers used to fabricate polymeric frameworks with a high thickness of cross-connecting synthetic bonds by a restoring cycle. The most commercially utilized thermoplastics and thermosetting along with their certain applications were represented in Table 5.1. Polyolefins are one of the elastomers that are highly popular due to their low density, cheap

TABLE 5.1
Commercially utilized thermoplastics and thermosetting with popular applications (Diaz and Phan 2016, Singh et al., 2017, Dodiuk & Goodman, 2014)

Thermosetting	Applications
Polyurethane (PU)	Medical devices, footwear, automobile components, freezers and refrigerators, bedding and furniture
Bakelite	Insulating material in various equipment and household utensils, electrical systems
Epoxy	Adhesive and coating substance, potting
Polyester	Staple fiber, bottles of beer and juices, adhesives
Vinyl ester	Laminating process, vessels and tanks, marine industry
Phenol formaldehyde	Adhesive and coating substance, fiberglass clothes, circuit boards, billiard balls
Silicon	Electrical and thermal insulations, adhesive, kitchen and medical utensils
Melamine	Kitchen utensils, dinnerware, laminating flooring
Thermoplastics	**Applications**
Poly carbonate (PC)	Electronic and telecommunications appliances
Polypropylene (PP)	Clear bags, food packaging, mats and carpet
Polyamides (PA)	Toothbrushes, textile industry, cams, gears, ropes
Poly ethylene terephthalate (PET)	Wires, beverage bottles, food packaging, staple fiber
High density poly ethylene (HDPE)	Automotive parts, tubes, buckets, cable and wire insulations, pipes
Low density poly ethylene (LDPE)	Screen cards, plastic bags, food containers, trays, general packaging films
Polystyrene (PS)	Thermal insulations, food packaging, toys, electronic appliances, dinnerware, CD cases
Poly vinyl chloride (PVC)	Serum and blood bags in medical field, automobile parts, fittings and tubes, medicine and food packaging
Poly lactic acid (PLA)	Disposable garments, hygiene products, bags, cups, medical implants
Acrylonitrile butadiene styrene (ABS)	Automotive components, electronic components, musical instruments, drain pipes

production, and relatively low cost for multiple applications (da Silva & Wiebeck, 2020). Irrespective of a large number of available polymers, the PW found on earth mainly consists of five thermoplastics: polystyrene, low- and high-density poly ethylene (available in packaging material form), polypropylene, poly ethylene terephthalate (Diaz and Phan 2016, Eriksen et al., 2018). For managing the PW and protecting the environment, plastic recycling becomes the leading issue worldwide.

5.1.1 Recycling Techniques

With an aim to reduce the generation of PW and its harmful impact on a socio-environment, a strategy based on 3R (reduce, reuse, and recycle) has been adopted by the developed and various developing countries. As per the 3R approach for PW management, any product irrespective of its materials should follow these guidelines during the entire life cycle: (i) non-generation of waste, (ii) reduction of waste, (iii) reuse, (iv) recycling, (v) landfilling (treated as last option). Recycling basically involves a process in which a waste material is reutilized for fabricating any product or can be reused in any useful application by enhancing or transforming its properties and other characteristics. As per the literature survey, polymer recycling is broadly classified into the following four categories (Al-Salem et al., 2017, Spinacé & De Paoli, 2005, Hamad et al., 2013, Cruz et al., 2011).

Primary (1°) recycling: This type of recycling involves sorting plastic waste from high-purity or semi-clean PW. It is generally preferred by industries to reduce their own PW by separating the easily identifiable clean scrap from the contaminated parts. Sometimes, the clean scrap is added into collected waste to improve their properties as well as performance. More often the PW collected from household applications is treated by 1° recycling. This type of recycling is particularly helpful in reducing the emission of toxic gases such as sulfur dioxide, nitrogen oxide, and carbon dioxide (Al-Salem et al., 2009). Since the municipal solid waste contains highly contaminated PW, it restricts the applicability of this particular recycling technique in such applications.

Secondary (2°) recycling: This type of recycling is particularly accomplished by mechanical means and hence it is also called mechanical recycling. It is the widely accepted technique for recycling of large-scale PW in the world due to the following reasons (Garcia & Robertson, 2017):

- lower infrastructure cost for sorting and treating the PW as compared to other recycling techniques.
- low operational cost due to already established and existing standards along with equipment for processing the PW.

The presence of multiple polymers along with impurities in the PW hampers their reusability and thus, sorting and separation steps become essential before their processing (Al-Salem et al., 2009). The existence of different polymers in the PW causes phase partition and results in development of compatibility issues, thus adversely affecting the mechanical performance of the newly fabricated component.

Moreover, it offers significant challenges during the processing of PW. Impurities (like inks and particulates) limit the creation of superior grade and homogeneous final recyclates (da Silva, 2016, da Silva et al., 2018). So, it becomes necessary to develop highly advanced techniques with an aim to separate the postconsumer polymers from different existing materials as far as possible. Apart from this, the thermomechanical degradation occurred during the processing of PW imparts major challenges. The recycling of PW is restricted by the frequency of cycles that the material can withstand without affecting its performance/properties beyond a specified level (Rahimi & García, 2017).

Two-degreerecycling involves the following steps during the processing of PW:

- Separation of PW: The first step involves sorting and separation of different materials from the collected PW. Different polymers are segregated on the basis of their chemical composition, colour, density, or any other specific property.
- Form transformation: The next step after separation is to transform the collected PW in the form of small flakes, powder, or granules through rotary mills.
- Wash: Chemical washing needs to be performed for removing glue and organic and inorganic compounds from the polymer. Mostly, caustic soda and surfactants are commonly used for cleaning and washing purpose.
- Drying: It is an important step for reducing degradation of polymers during mechanical recycling. For example, polymers (polyesters or polyamides) are susceptible to hydrolysis as they are obtained by polycondensation.
- Agglutination: The next step is to reduce the volume of the binder to ensure its homogenization after washing and drying. This encourages the last cycle that is completed in an extruder.
- Reprocessing: The last step is transformation of polymers in the form of polymer grains. After this, the grains are packed and sent to industries for manufacturing new parts.

Tertiary (3°) recycling: In this type of recycling, monomers of different polymeric materials are recovered from the PW through the depolymerization process (Singh et al., 2017). In other words, through this recycling the basic raw materials required to form various plastics are generated and hence it contributes largely towards the energy sustainability as polymers are basically petroleum-based products (Chawla et al., 2020, Kumar et al., 2011). This type of recycling is also called chemical recycling. Depolymerization of PW can be performed either chemically (solvolysis) or through thermal (thermolysis) means. Further, the process can be executed in the absence of air (pyrolysis) or under controlled conditions (gasification).

Quaternary (4°) recycling: This type of recycling is particularly performed on the PW that cannot be recycled, such as waste generated from medical applications and packing of hazardous products and is mostly suitable for thermosetting polymers (Ignatyev et al., 2014). After over and over reusing, a polymer begins to lose its properties and ultimately leads to landfilling that contaminates the earth's surface as well as generates toxic gases. With the advancements of new incinerators,

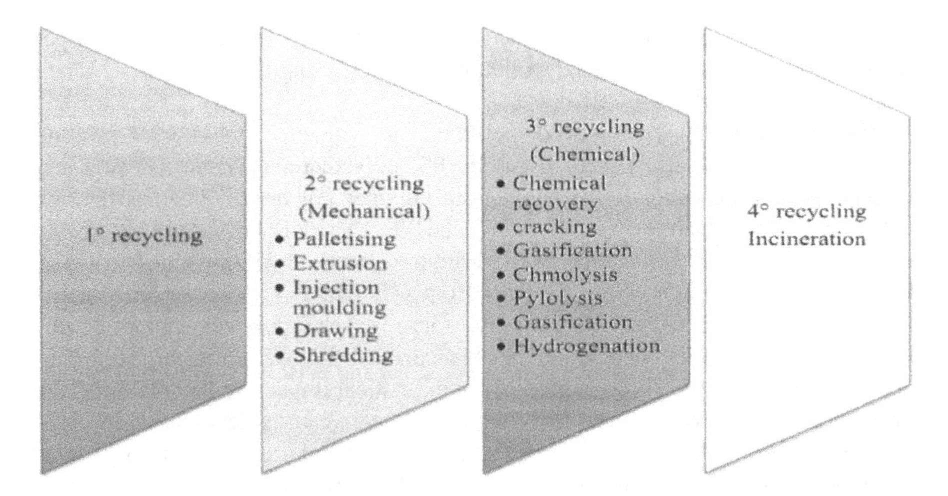

FIGURE 5.1 Recycling techniques of PW.

numerous analysts proposed that by incineration (burning) energy can be recuperated from PW, which leads to reduction in waste; thus, 4° recycling is also called a energy recovery technique. Thermal and electrical energy is generated from the energy (chemical) present in the bonds of a long chain structure of PW. During incineration, plastics generally produces gases like CO_2, SOx, and NOx as they are derived from crude oils [Rezaei et al., 2015, Sun et al., 2016]. There are a few techniques used to control these harmful pollutant particles, which include (i) filtration, (ii) addition of ammonia in combustion chamber, (iii) flue gas cooling, (iv) acid neutralization, and (v) flue gas cooling (Rahimi & García, 2017, Chawla et al. 2018). The previously discussed recycling techniques are summarized in Figure 5.1.

5.1.2 Recycled Polymers with Reinforced Materials

Competition among the manufacturing industries forces them to reduce the cost of their products. The usage of recycled material is quite helpful in achieving this goal and at the same time it reduces the industrial and municipal waste also. Chemical recycling (pyrolysis) of PW results in various environmental hazards and landfilling generates greenhouse gases. So, mechanical/2° recycling comes out as an only solution to manage/reduce the continuously increasing PW. The new advanced technologies to process the PW with cost-effective solutions needs to be developed urgently. Additive manufacturing (AM), one of the relatively new technologies, became quite popular in a very short time due to its capability to fabricate complex features and it is expected to rise with 23% market growth by 2021 compared to 2016 (Shah et al., 2019). AM techniques such as 3D printing and fused deposition modeling (FDM) have the capability to utilize filament prepared from waste materials and thus offers a potential solution to reduce the PW generated from different applications (Mikula et al., 2020). Researchers proposed to utilize recycled thermoplastics in an open-source 3D printer/FDM setup by preparing the filament through plastic extruders (Baechler et al., 2013). The FDM process is being given continuous attention to utilize recycled plastic materials with the use of proper reinforcements for cement-based building materials (Singh et al., 2016a, Singh et al., 2016b, Farina et al., 2016).

The most commonly employed thermoplastic filaments in FDM/3D printing are PLA, PC, ABS, LDPE, HDPE, PET, PS, polyetheretherketone (PEEK) and these filaments are generally prepared from polymer powder or granules through extrusion process. Currently, ABS, LDPE, HDPE, PP, PET, PS, and PC are being recycled globally for fabrication of 3D-printing filaments. Throughout the world, various researchers fabricated filaments from different PWs with or without the additives, as represented in Table 5.2. But these recycled polymers cannot be utilized directly due to their poor mechanical and physicochemical properties (Lanzotti et al., 2019). These properties have a critical impact on the quality of the printed product. Also, multiple recycling of polymers causes degradation in properties (such as break strength, viscosity, molecular weight, etc.) due to high temperature and shear stresses that occurred during the extrusion process and ultimately affects the quality of the extruded products.

Most of the commercially available 3D printers utilize ABS and PLA as filaments. Although ABS is toxic in nature but it exhibits enormous applications (particularly in electronic and automobile industry) due to its high corrosion resistance and impact strength. On the other hand, PLA is bio-compatible and biodegradable polymer but highly sensitive to elevated temperatures (Duval, 2014). Researchers reported 30% and 60% degradation in PLA after recycling of three and seven times, respectively (Brüster et al., 2016). Some of the researchers studied the effect of multiple recycling of PLA polymers and reported that the polymer chain length reduces as the injection cycles increase, which ultimately increases the melt flow rate (rheological property) and decreases the various mechanical properties (break tensile strength and break tensile strain) (Sanchez et al. 2015). It is a well-established fact that multiple recycling of PLA has an adverse effect on mechanical properties and viscosity of polymer which prohibits its further utilization for 3D-printing applications (Anderson, 2017, Zhao et al., 2018a,b). As mechanical strength of the PLA reduces significantly with the recycling cycles, it was proposed to apply polydopamine (PDA) coating on the recycled polymer filaments (Zhao et al., 2018b). Some researchers suggested adding virgin PLA to the shredded and recycled PLA filament in order to improve the thermal and mechanical properties along with the viscosity of the blend (Zhao et al., 2018a,b).

A lot of researchers reinforced materials like carbon nanotubes, nano-particles, metallic powders, fibres (glass, jute, sisal, coconut, carbon), wood, bamboo, etc. in mechanically recycled (1° and 2°) thermoplastics and a thermosetting with an aim to induce or enhance specific properties (Adhikary et al., 2008, Jen & Huang, 2014, Lu & Oza, 2013, Oliveux et al., 2015, Onwudili et al., 2016, Yildirir et al., 2015). Pan et al. (2018) reported that the addition of a 1% mix of nano-metallic powders (Fe-Si-Al or Fe-Si-Cr) in recycled HDPE and PP filaments improves the Young's modulus and yield strength significantly. Further, the authors identified that the probability of cracks formation in a base matrix reduces with the introduction of metallic powders. In another study, the introduction of SiC/Al_2O_3 in recycled HDPE with paraffin wax as a binder dramatically increases the mechanical strength

TABLE 5.2
Fabricated 3D-printed filaments through reuse of different PWs

Study	Base Matrix	Additives	Source of waste
Woern et al. (2018)	ABS	–	ABS-post consumer
Mohammed et al. (2017)	ABS	–	Failed 3D prints and virgin pellet material
Chawla et al. (2020)	ABS	–	Unknown
Chawla et al. (2020)	ABS	Bakelite, wood dust	Unknown
Sahajwalla & Gaikwad (2018)	PC	–	Electronic waste
Hart et al. (2018)	LDPE/LLDPE	–	Meal bags
Andersson et al. (2004)	LDPE/LLDPE	–	LDPE, CA 8200LLDPE, LE 1000
Woern et al. (2018)	PP		McDonnough plastics
Pan et al. (2018)	PP		Postconsumer hard plastics
Brouwer et al. (2018)	Mixture of PP, PE, PET		Household plastic products
Zander et al. (2019)	Mixture of PP, PS, PET	–	Recycling bins
Zander et al. (2018)	PET	–	Salad containers, water bottles
Idrees et al. (2018)	PET	Biochar	Water bottles
Exconde et al. (2019)	PET	–	Unknown
Chong et al. (2017)	HDPE	–	Milk, shampoo and household bottles
Baechler et al. (2013)	HDPE	–	Unknown
Pan et al. (2018)	HDPE	Silicon, Chromium, Aluminum	Plastic bags
Singh et al. (2019b)	HDPE	Zirconium oxide	Unknown
Cisneros-López et al. (2020)	PLA	–	Industrial waste mix
Zhao et al. (2018b)	PLA	Polydopamine (PDA)	3D printed broken PLA parts
Anderson (2017)	PLA	–	Unknown
Gkartzou et al. (2017)	PLA	Craft lignin	INGEO 2003D (Natureworks LLC)
Tian et al. (2016a, b)	PLA	Carbon fiber reinforced (CFR)	Filament (FLASHFORGE Corp Japan)
Woern et al. (2018)	PLA	–	Failed 3D prints

of filaments. Some authors reported improved dimensional stability and mechanical properties when wood fibres were reinforced in plastics (Adhikary et al., 2008, Singh & Singh, 2015). The bonding between the base matrix and the reinforced material plays an important role in order to achieve the desired properties. In order to improve the bonding between the base matrix and the reinforced material, certain bonding agents such as silanes, organo-titanes, maleated polypropylene, etc were employed by different researchers (Adhikary et al., 2008).

Gkartzou et al. (2017) reported that the addition of a lignin biopolymer reduces the Young's modulus and tensile strength of a PLA specimen by 6% and 18%, respectively, whereas melting properties improve compared to pure PLA samples. Idrees et al. (2018) highlighted that a higher thermal resistance along with improved tensile strength (32%) and modulus of elasticity (60%) of PET materials were obtained when biocarbon (<100 μm) was added into it. Singh et al. (2019) prepared a composite filament by reinforcing zirconium oxide in recycled HDPE and investigated that the coefficient of friction of composite filaments was 40% lower than those of unreinforced filaments. The authors proposed the composite material to be used for low-temperature bearing applications. The work performed by the various researchers for composite filament fabrication along with obtained mechanical properties has been summarized in Table 5.3.

TABLE 5.3

Fabricated composite filaments with properties for 3D printing applications

Base Matrix	Reinforced material with %	Tensile Modulus (GPa)	Tensile Strength (MPa)	Flexural Modulus (GPa)	Flexural Strength (MPa)	Reference
ABS	CF (10%)	1.67	32.78	–	–	Karaman & Çolak (2020)
ABS	CF (12%)	2.72	31.70	–	43.80	Patan (2019)
ABS	CF (7.5%)	2.5	41.5	–	–	Ning et al. (2015)
	CF (15%)	2.25	35	–	–	
ABS	CF (10%)	7.7	52	–	–	Zhong et al. (2001)
	GF (18%)	–	58.6	–	–	
ABS	CF (20%)	8.4	66.8	–	–	Hill et al. (2016)
ABS	CF (13%)	8.15	53	–	–	Kunc (2015)
ABS	Al (1.5%)	2.55	44.56	–	–	Çantı (2016)
	MWCNT (1.5%)	1.68	41.2	–	–	
	SiO_2 (1.5%)	2.49	44.72	–	–	
	ZrB_2 (1.5%)	2.14	40.2	–	–	
ABS	C (1.6%)	–	43	–	–	Mori et al (2014)
ABS	Fe (10%)	0.96	43.4	–	–	Hwang et al (2015)
	Fe (40%)	0.95	36.2	–	–	
	Cu (30%)	0.91	26.5	–	–	
ABS	GF (20%)	5.7	54.3	–	–	Duty et al. (2015)
	GF (40%)	10.8	51.2	–	–	
ABS	Jute fiber (5%)	1.54	25.9	–	–	Perez et al. (2014)
PLA	Talc (2%)	1.47	57.9	–	–	Gao et al. (2019)
	CF (5%)	1.10	31.7	–	–	

TABLE 5.3 (Continued)
Fabricated composite filaments with properties for 3D printing applications

Base Matrix	Reinforced material with %	Tensile Modulus (GPa)	Tensile Strength (MPa)	Flexural Modulus (GPa)	Flexural Strength (MPa)	Reference
PLA	CF (15%)	7.54	53.4	–	–	Ferreira et al. (2017)
PLA	CF (20%)	4.54	29.96	–	–	Hodzic & Pandzic (2019)
PET-G	CF (20%)	4.26	32.79	–	–	
PLA	CF	0.74	41.3	2.93	75.6	Liu et al. (2019)
	Al	0.83	51.1	3.27	97.8	
	Ceramic	1.05	46.5	4.62	100.1	
PLA	Cu (10%)	0.93	42	–	–	Tian et al. (2016a, b)
	CF (27%)	–	–	30	355	
PLA	GF (0%)	0.71	44.75	2.52	75.50	Rajpurohit & Dave (2019)
	GF (1%)	0.81	43.65	2.25	56.65	
	GF (3%)	0.60	31.60	2.34	61.80	
	GF (5%)	0.57	24.65	2.11	50.55	
PLA	C (34%)	23.8	91	–	–	Li et al. (2016)
Nylon	C (6.6%)	19.5	185.2	–	–	Van Der Klift et al. (2016)
	C (6.9%)	14	140	–	–	
	C (20.7%)	35.7	464.4	–	–	
Nylon	Aramid (2%)	–	51.45	–	98.16	Nagendra & Prasad (2020)
Nylon	Kevlar (4.04%)	1.77	31	–	–	Melenka et al. (2016)
	Kevlar (8.08%)	6.92	60	–	–	
	Kevlar (10.1%)	9	84	–	–	
Nylon	AF (8%)	4.23	110	–	–	Dickson et al. (2017)
	AF (10%)	4.76	161	–	–	
	GF (8%)	3.29	156	–	–	
	GF (10%)	4.91	212	–	–	
PPS	CF (50%)	26.4	92.2	–	–	DeNardo (2016)
PEI	CNT (4.7%)	3	125.3	–	–	Gardner et al. (2016)
PE	Cu (25%)	0.7	17.12	–	–	Nabipour et al. (2020)
	Cu (50%)	0.79	18.25	–	–	
	Cu (75%)	1.2	19.41	–	–	
HDPE	Cardboard (20%)	0.32	9.04	–	–	Gregor-Svetec et al. (2020)
	Cardboard (50%)	0.12	2.05	–	–	
	Cardboard (75%)	0.10	1.89	–	–	

Taking into account the importance and recent advances in the field of recycling, countless surveys have just examined the problem and solution of polymer waste. But much less work has been done on 2° recycled thermoplastic polymers for preparation of sustainable structures (Simon et al. 2018, Unnisa & Hassanpour, 2017, Silveira et al., 2018, Garcia & Robertson, 2017, Al-Salem et al., 2017, Butler et al. 2011, Ignatyev et al., 2014, Al-Salem et al., 2010, Al-Salem et al., 2009, Spinacé & De Paoli, 2005, Hamad et al., 2013). Also, the mechanical behaviour of 3D-printed structures with multiple materials (multi-layer printing) has not been extensively explored. So, in this research work, an effort has been performed to print multi-layered specimens fabricated through 2° recycled ABS thermoplastic with the reinforcement of bakelite powder, Fe powder, and wood dust. The particular reinforced materials were selected, keeping in mind that bakelite powder can improve thermal properties (heat carrying capacity) of ABS, wood dust can impart insulating properties, whereas Fe powder is for magnetic properties. The structure/tiles printed by these multi-layered composite filaments can be utilized in cold regions for keeping the heat inside the confined space.

5.2 CASE STUDY: SUSTAINABLE STRUCTURE FABRICATED THROUGH FDM FROM PW

In this case study, all waste materials (base matrix as well as reinforcements) were utilized to fabricate the composite filaments. First of all, the base material i.e. 2° recycled ABS purchased from a local vendor was placed in a vacuum oven for five hours at 50°C. This process was carried out to remove the traces of any moisture that may present in the ABS granules. The waste Fe powder and wood dust were collected from the institute's (Lovely Professional University) machine shop and carpentry shop, respectively. The waste bakelite (solid form) obtained from the local electrical industry was transformed into powder form by cryogenic crushing performed at –196°C. In order to ensure uniformity, the grain size of 50 μm was selected for all the reinforced materials. In the next step, the melt flow index (MFI) as a rheological property determines the flow behaviour of the plastic and was calculated under standard conditions as per ASTM D1238 standard. The variation in MFI with a percentage of reinforcement for all the composite filaments was also investigated and reported in Table 5.4. The MFI was measured thrice for all the selected compositions of reinforced materials in order to minimize the randomness and the average of the three readings was considered as a final output.

TABLE 5.4
Variation in MFI of 2° recycled ABS with reinforcements percentage

	Base matrix and reinforced material	Reinforcement percentage				
		0%	2.5%	5%	7.5%	10%
MFI (g/10min)	2° recycled ABS	16.32	–	–	–	–
	2° recycled ABS + Fe powder	–	20.32	22.92	19.96	18.38
	2° recycled ABS + bakelite powder	–	21.26	18.28	16.24	14.76
	2° recycled ABS + wood dust	–	21.50	27.68	18.24	14.10

The observations indicated that the MFI of selected composites (in general) increases up to a certain percentage of reinforced material and then it begins to reduce with further addition of reinforcement. In this study, reinforcement percentage was kept limited to 10% as nozzle clogging was observed with higher levels. After measuring the MFI, three composite filaments were decided to be prepared by taking 10% of the reinforcement as this level offers a maximum utilization of the industrial waste. Also, the value of MFI with 10% of reinforcement in all three cases was close to the MFI of recycled ABS without any reinforcement and hence it will not create complications during printing of the specimens. The composite filaments were prepared by varying the processing (extrusion) conditions of speed/rpm, temperature, and load on the twin-screw extruder (TSE). In order to reduce the number of experiments, the L9 orthogonal array has been utilized to plan the experiments. The processing levels were decided by performing the number of pilots run and the final experimentation for preparation of composite filaments was executed as per Table 5.5. The methodology to fabricate composite filaments has been described in Figure 5.2.

TABLE 5.5
Processing conditions for filaments fabrication

Experiment No.	1	2	3	4	5	6	7	8	9
Load (kg)	10	10	10	12.5	12.5	12.5	15	15	15
Temperature (°C)	225	235	245	225	235	245	225	235	245
Speed (rpm)	70	80	90	80	90	70	90	70	80

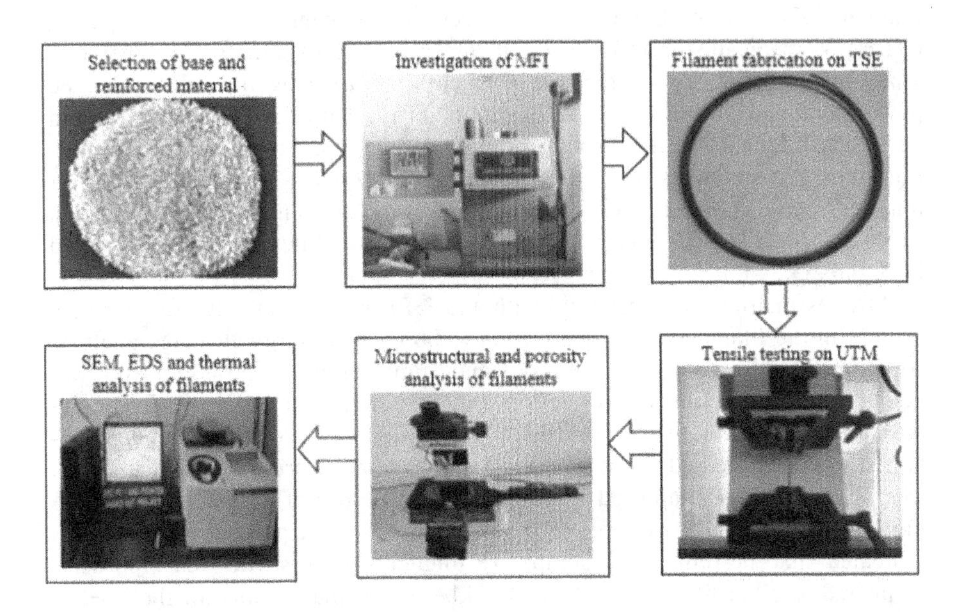

FIGURE 5.2 Methodology employed in research work.

TABLE 5.6
Mechanical properties of fabricated composite filaments

	Bakelite-reinforced filaments		Wood dust–reinforced filaments		Fe powder–reinforced filaments	
S. No.	% BE	PS (MPa)	% BE	PS (MPa)	% BE	PS (MPa)
1	5.3 ± 0.7	30.82 ± 1.7	6.0 ± 0.7	25.66 ± 1.4	3.17 ± 1.5	32.20 ± 1.1
2	4.2 ± 0.5	28.36 ± .1.5	4.1 ± 0.4	24.46 ± 1.5	3.85 ± 1.1	30.37 ± 1.3
3	3.8 ± 0.4	23.85 ± 0.9	3.8 ± 0.5	21.60 ± 1.1	3.98 ± 1.2	28.99 ± 1.5
4	3.7 ± 0.4	29.10 ± 1.4	3.7 ± 0.6	23.84 ± 1.3	3.18 ± 1.5	35.43 ± 1.3
5	3.1 ± 0.3	26.33 ± 1.2	3.1 ± 0.3	20.15 ± 1.0	4.45 ± 1.0	28.51 ± 0.8
6	4.9 ± 0.6	29.80 ± 1.4	5.7 ± 0.5	24.13 ± 1.2	4.32 ± 1.1	30.77 ± 1.4
7	2.7 ± 0.3	27.20 ± 1.3	3.4 ± 0.4	18.84 ± 0.8	3.87 ± 1.2	29.67 ± 1.0
8	5.0 ± 0.5	27.82 ± 1.6	5.4 ± 0.6	22.66 ± 1.3	3.82 ± 1.3	31.34 ± 1.5
9	4.4 ± 0.4	28.23 ± 1.3	3.3 ± 0.4	20.21 ± 1.2	3.12 ± 1.4	26.26 ± 1.2

The next step after preparing the composite filaments was to investigate their mechanical properties. In this regard, two important mechanical properties (peak strength and percentage break elongation) of the composite filaments were determined by tensile testing performed on the universal testing machine (UTM). The results obtained for all the composite filaments are represented in Table 5.6. The observations indicated that Fe powder–reinforced composite filaments exhibit a maximum peak strength (PS), whereas wood dust–reinforced filaments show a minimum PS. However, the percentage break elongation (%BE) of wood dust–reinforced filaments was highest among the other two types of filaments. The composite filaments prepared by the reinforcing Fe powder indicate a minimum %BE due to the brittle nature of reinforcement and hence the filaments break without indicating any yielding (ductile behaviour). Based on the obtained data, the modulus of elasticity (E) and modulus of toughness (T) were calculated for the composite filaments and are shown in Table 5.7. Stress-strain curves for the three composite filaments are represented in Figures 5.3–5.5.

After estimating the tensile strength and %BE of the filaments, the porosity present in the filaments was analyzed at ×100 by metallurgical image analysis software (MIAS) installed in the tool maker microscope. The composite filaments were cut at the cross section and then polished before the porosity measurement as per the ASTM B276 standard. The porosity percentage obtained in the fabricated composite filaments have been presented in Table 5.8. The porosity images captured by software for all types of filaments having the highest and lowest PS are represented in Figure 5.6. The porosity observations indicated that Fe-reinforced composite filaments possess the least porosity compared to other filaments, which could be one of the reasons for their better peak strength. The filaments reinforced with bakelite and wood dust exhibit less

TABLE 5.7
Modulus of toughness (T) and modulus of elasticity (E) for composite filaments

S. No.	Bakelite-reinforced filaments		Wood dust–reinforced filaments		Fe powder–reinforced filaments	
	T (MPa)	E (GPa)	T (MPa)	E (GPa)	T (MPa)	E (GPa)
1	0.835 ± 0.3	0.695 ± 0.15	0.697 ± 0.2	0.477 ± 0.1	0.46 ± 1.2	1.132 ± 1.2
2	0.544 ± 0.2	0.746 ± 0.25	0.475 ± 0.15	0.772 ± 0.25	0.50 ± 1.5	0.871 ± 1.1
3	0.408 ± 0.15	0.636 ± 0.2	0.369 ± 0.15	0.758 ± 0.2	0.45 ± 1.0	0.832 ± 1.5
4	0.467 ± 0.2	0.919 ± 0.35	0.679 ± 0.25	0.396 ± 0.1	1.10 ± 1.4	0.486 ± 1.2
5	0.539 ± 0.25	0.924 ± 0.4	0.255 ± 0.1	0.711 ± 0.2	0.57 ± 1.1	0.818 ± 1.4
6	0.525 ± 0.2	0.917 ± 0.3	0.378 ± 0.2	0.862 ± 0.25	0.60 ± 0.8	1.079 ± 0.8
7	0.615 ± 0.25	0.781 ± 03	0.372 ± 0.15	0.632 ± 0.15	0.53 ± 1.1	1.041 ± 1.0
8	0.736 ± 0.35	0.573 ± 0.2	0.232 ± 0.1	0.894 ± 0.2	0.83 ± 1.2	0.582 ± 1.5
9	0.252 ± 0.15	1.323 ± 0.4	0.230 ± 0.15	0.836 ± 0.25	0.42 ± 1.1	0.921 ± 1.2

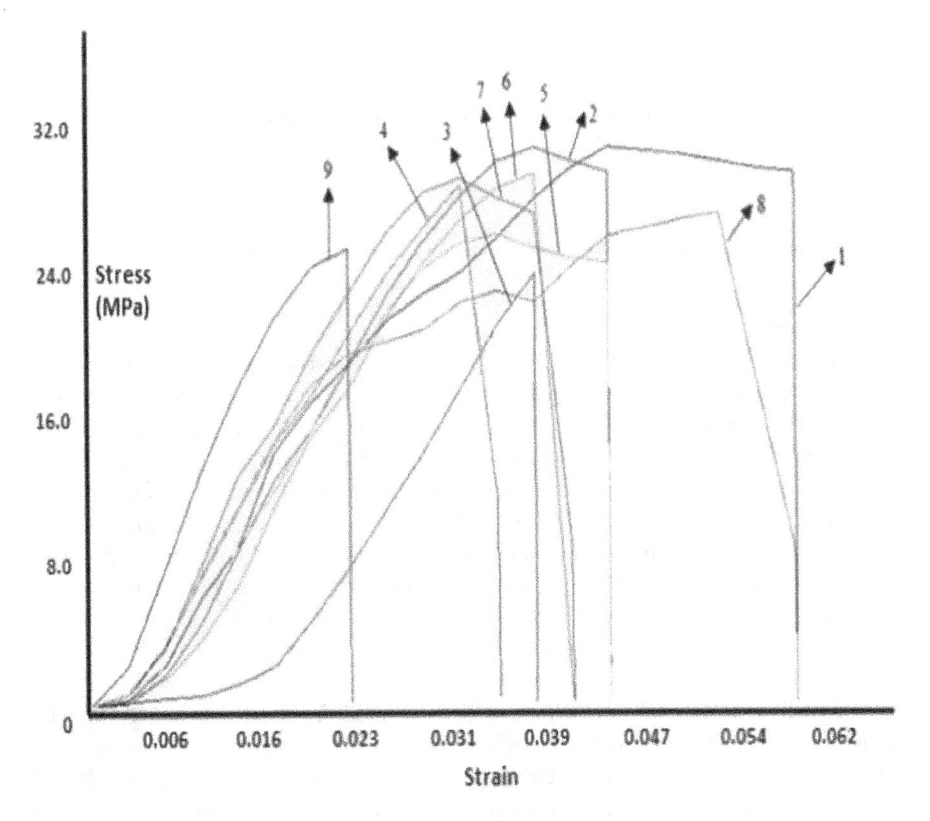

FIGURE 5.3 Stress-strain curve for bakelite powder–reinforced composite filaments.

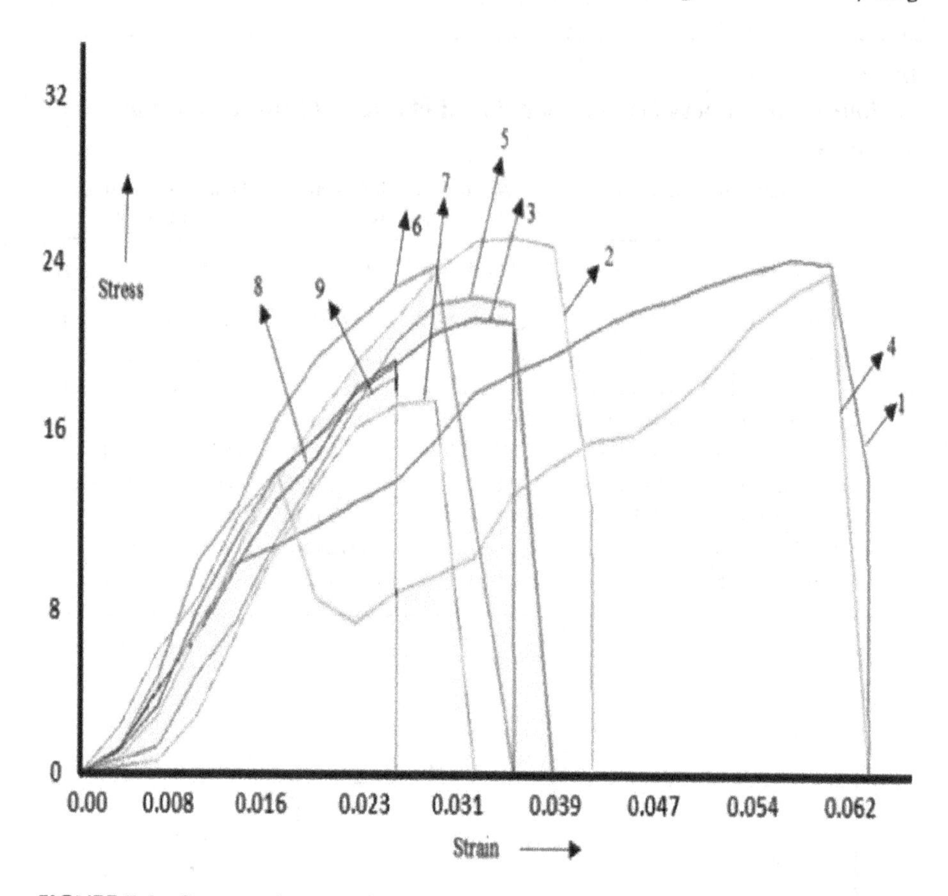

FIGURE 5.4 Stress-strain curve for wood dust–reinforced composite filaments.

porosity particularly at an extrusion temperature of 225°C, whereas a temperature of 235°C yields a minimum porosity in Fe-reinforced filaments. This may be due to a higher specific heat carrying capacity of Fe powder compared to other two reinforcements. The particle shape of the reinforcements also influences the porosity as 2D structured wood dust particles occupied less volume in an ABS matrix compared to spherical-shaped bakelite and Fe powder. Due to this reason, the wood dust–reinforced filaments show a maximum porosity and hence minimum strength among the fabricated composite filaments. The higher porosity results in poor mechanical behaviour of the fabricated filaments and vice versa. The particle shape of the reinforcements was distinguished through scanning electron microscopic (SEM) images, as shown in Figure 5.7.

A signal to noise (S/N) ratio has been calculated for the PS of the composite filaments with an aim to predict the effect of extrusion parameters. S/N ratios were calculated based on "larger is better" criterion and represented in Table 5.9. Based on Table 4.6, mean S/N ratios were plotted (refer to Figure 5.8) for each input parameter to investigate their effect on the PS of the filaments. Speed/rpm of the

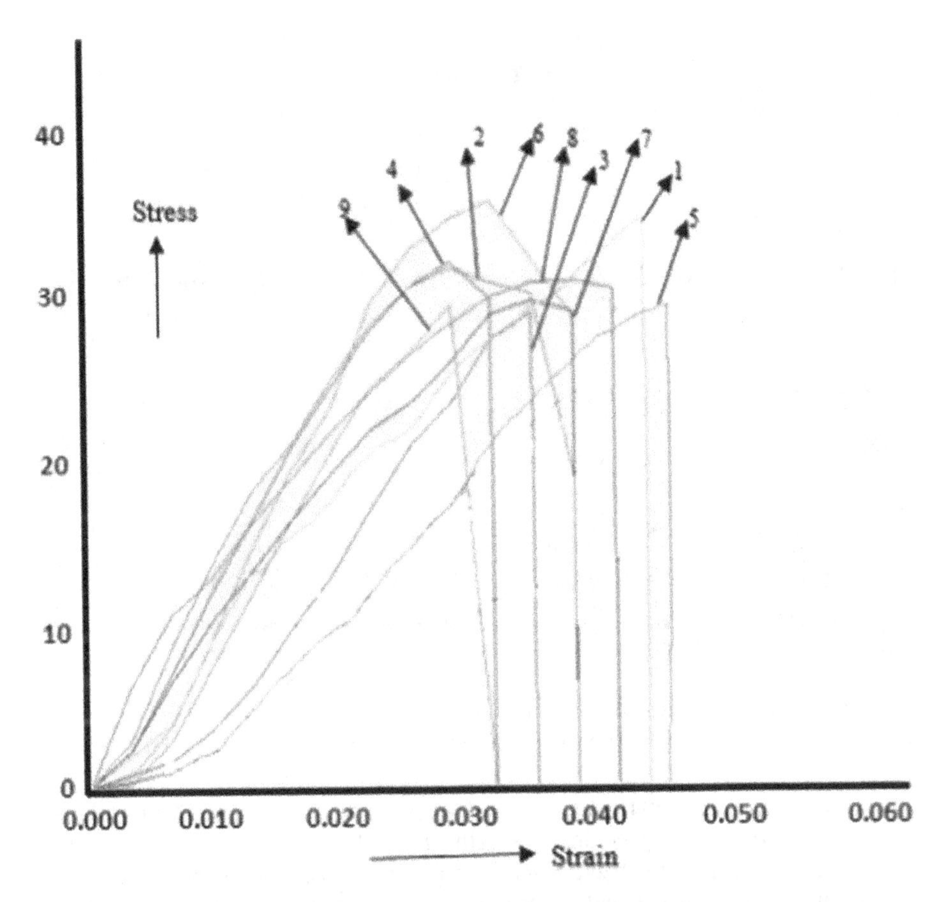

FIGURE 5.5 Stress-strain curve for Fe powder–reinforced composite filaments.

TABLE 5.8
Porosity percentage in composite filaments

	Porosity Percentage		
Experiment No.	**Bakelite powder–reinforced filament**	**Wood dust–reinforced filament**	**Fe powder–reinforced filament**
1	5.32	3.64	4.96
2	7.66	11.52	5.32
3	12.64	29.76	7.66
4	6.56	4.96	3.11
5	13.70	15.87	8.74
6	12.24	34.07	4.12
7	13.79	6.18	7.31
8	12.00	10.00	3.64
9	14.55	32.38	9.20

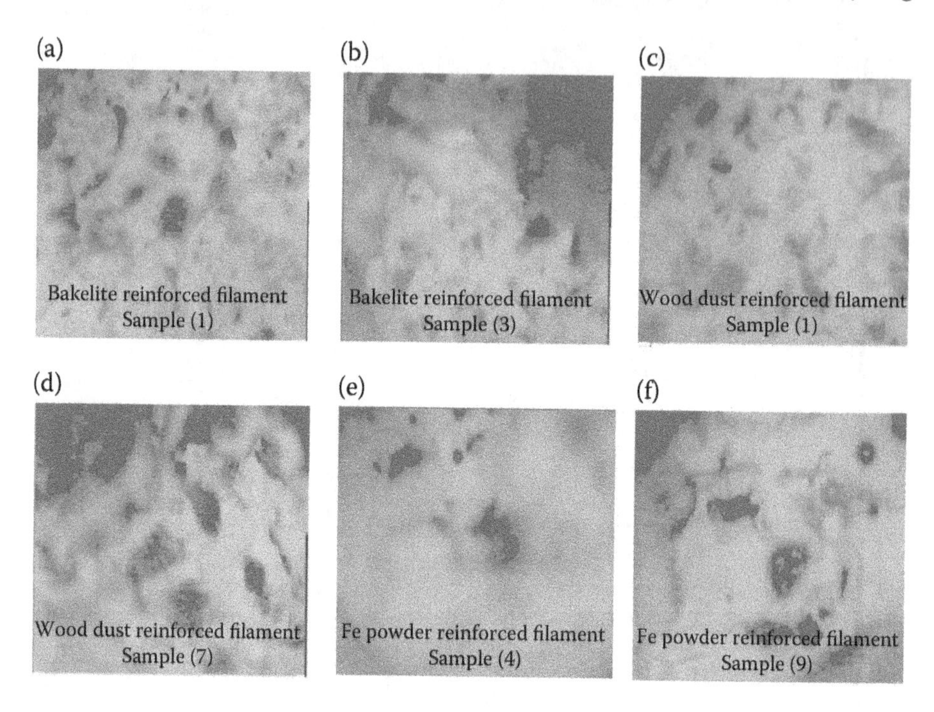

FIGURE 5.6 Porosity images for composite filaments reinforced with bakelite (a, b), wood dust (c, d), and Fe powder (e, f) having the highest and lowest PS, respectively.

twin-screw during the extrusion process affects the PS of the filaments predominantly and as the speed increases, the PS reduces dramatically for all the composite filaments. It has been observed that lower speed yields more PS due to the fact that both the matrix and reinforced materials get sufficient time for uniform blending compared to high speed. Similar to speed, extrusion temperature also has a significant impact on the PS and as the temperature rises, the PS starts reducing sharply, particularly in Fe powder–reinforced composite filaments. The extrusion load is an important parameter that helps in smooth flow of the filament from the extruder opening affects the PS of all the three filaments differently. With an increase in load the PS increases slightly for bakelite-reinforced filaments, reduces sharply for wood dust–reinforced filaments, and first increases and then reduces for Fe-reinforced filaments. The optimized levels that result in a maximum PS for composite filaments reinforced with bakelite (15 kg, 225°C, 70 rpm), wood dust (10 kg, 225°C, 70 rpm), and Fe powder (12.5 kg, 225°C, 70 rpm) were identified to form S/N ratio plots.

Based on the optimized levels of PS, different composite filaments were fabricated on the TSE in order to examine their feasibility for printing applications. The composite filaments were utilized on the FDM machine to print a multi-layered tensile specimen, as shown in Figure 5.9. The specimen was printed with three layers of each composite filament i.e. a total of nine layers were printed. The outer layers on two sides of the tensile specimen were printed with filament reinforced with bakelite and Fe powder, respectively, as they possess better PS compared to

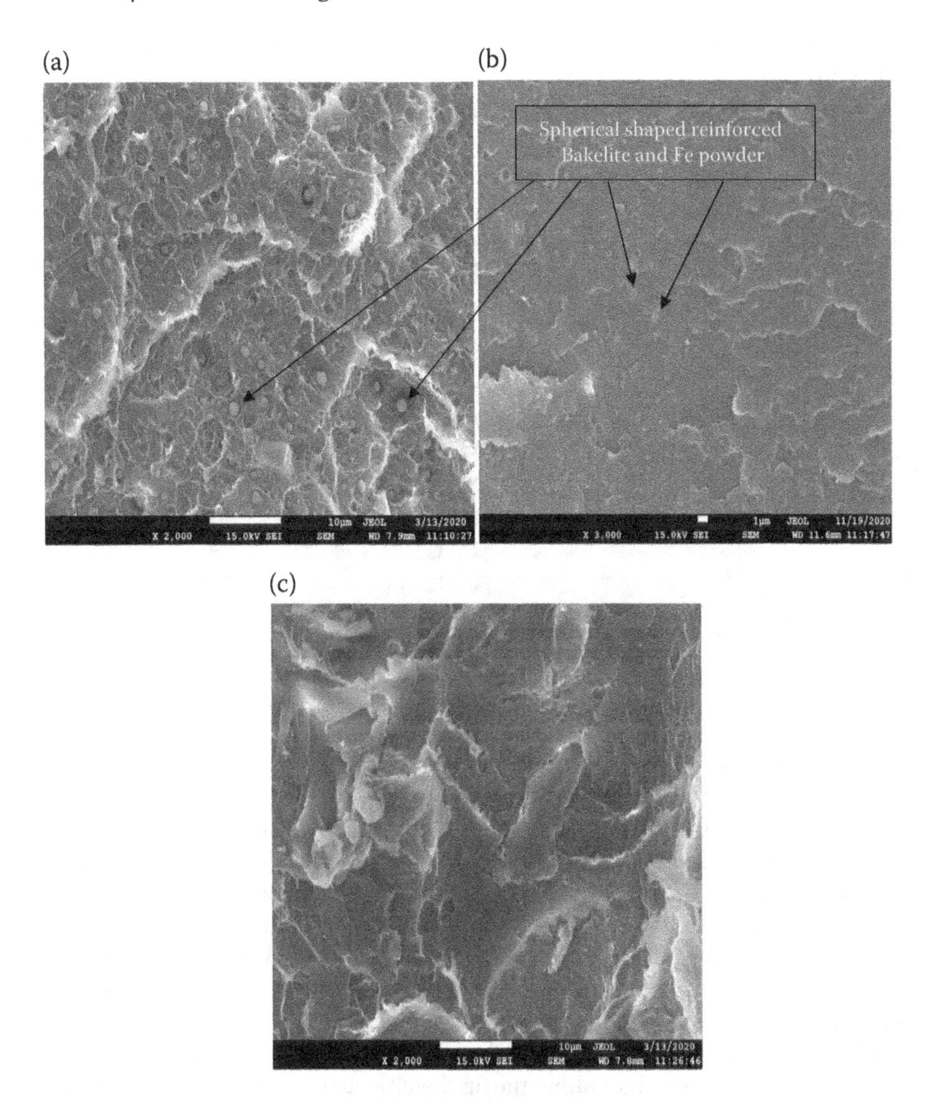

FIGURE 5.7 SEM characterization for composite filaments reinforced with a) bakelite, b) Fe powder, and c) wood dust.

wood dust–reinforced filaments. The inner layers were composed of wood dust–reinforced filaments. During the printing process, the composite filaments prepared from recycled materials does not offer any challenge. The specimens were printed by varying the FDM parameters like raster angle, infill density, and infill speed. After preparing the specimen, their strength was estimated by performing the tensile tests on UTM and the observed values were reported in Table 5.10. The present case study demonstrates the feasibility of utilizing 2° recycled ABS thermoplastic as filaments for preparation of sustainable structures through the FDM technique of additive manufacturing.

TABLE 5.9
S/N ratios for PS of composite filaments

	S/N ratios (dB)		
Experiment No.	**Bakelite powder–reinforced filament**	**Wood dust–reinforced filament**	**Fe powder–reinforced filament**
1	29.78	28.19	30.16
2	29.05	27.77	29.65
3	27.55	26.69	29.24
4	29.28	27.55	30.99
5	28.41	26.09	29.10
6	29.48	27.65	29.77
7	28.69	25.50	29.44
8	28.88	27.11	29.92
9	29.01	26.11	28.45

5.3 CONCLUSIONS

The repeated, long-chain, and stable polymeric structure of thermoplastics does not allow them to decompose in natural atmospheric conditions and thus causes environmental contamination along with various kinds of pollution (marine and terrestrial). Recycling of plastic waste becomes an essential and advantageous method to manage the waste and protect the environment from their harmful impacts. In this book chapter, the authors initially presented a comprehensive review on the various recycling techniques adopted by the various countries and concluded that mechanical recycling was observed to be the promising technique to process the polymeric wastes as high capital investment is required to establish plants for tertiary and quaternary recycling and does not make them feasible for industries, including the undeveloped countries. The polymeric waste processed through mechanical recycling has been successfully utilized in the form of filaments by various researchers for 3D printing and FDM applications with or without the reinforcements and thus giving a second life to waste plastics.

The authors of this book chapter also explored the feasibility to prepare a sustainable structure from secondary recycled ABS-based composite filaments reinforced with Fe powder, wood dust, and bakelite powder by multi-layer printing using the FDM technique as a case study. Before fabrication of filaments on the twin-screw extruder, the effect of reinforcement percentage on the rheological behaviour (MFI) was studied and it has been investigated that the MFI of the selected composites increases up to a certain percentage of reinforced material and then it begins to reduce with the further addition of reinforcement. The observations of tensile testing indicated that Fe powder–reinforced composite

(a)

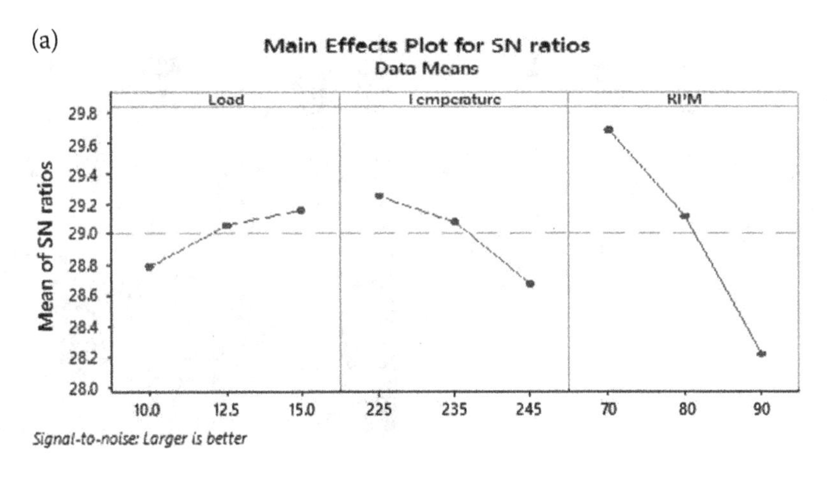

(b)

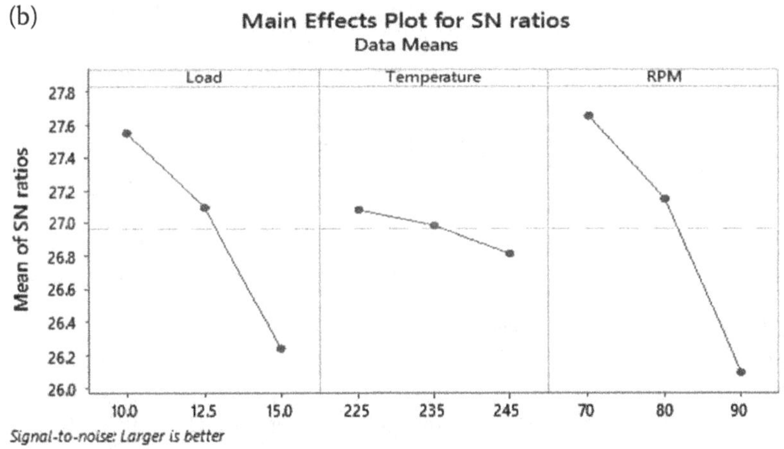

(c)

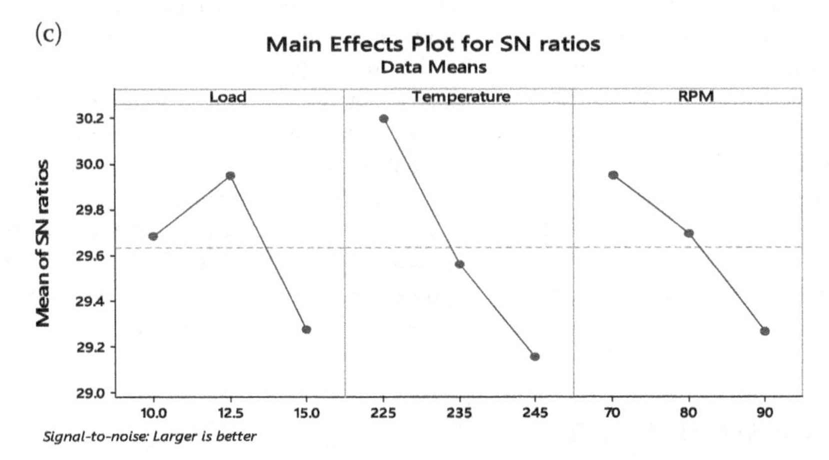

FIGURE 5.8 Mean S/N ratio plots for PS of composite filaments reinforced with a) bakelite, b) wood dust, and c) Fe powder.

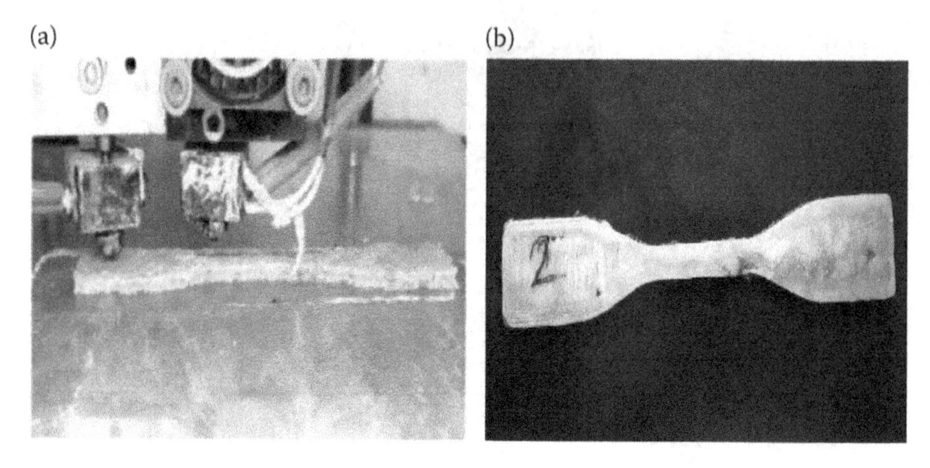

FIGURE 5.9 Multi-layered tensile specimen a) during and b) after printing on FDM.

TABLE 5.10
Strength of tensile specimen fabricated through FDM

Trial No.	1	2	3	4	5	6	7	8	9
PS (MPa)	11.85	13.84	15.35	12.45	11.08	13.10	8.96	9.98	6.68
BS (MPa)	10.5	11.84	10.07	8.07	10.66	6.68	7.45	9.98	5.10

PS: Peak Strength, BS: Break Strength.

filaments exhibit maximum peak strength, whereas wood dust–reinforced filaments show minimum peak strength. However, the percentage break elongation of wood dust–reinforced filaments was highest among the other two types of filaments. The mechanical behaviour of the filaments was also analyzed through porosity and scanning electron microscopic characterization. The optimized levels that result in maximum peak strength for composite filaments reinforced with bakelite (15 kg, 225°C, 70 rpm), wood dust (10 kg, 225°C, 70 rpm), and Fe powder (12.5 kg, 225°C, 70 rpm) were identified form S/N ratio plots. The filaments fabricated at optimized levels were then tested on the FDM to examine their feasibility for printing applications. The tensile specimen fabricated from recycled composite filaments by multi-layer printing possesses acceptable strength and does not offer any critical challenge during printing and thus can be used for printing of sustainable structures.

ACKNOWLEDGEMENTS

The authors are obliged to Manufacturing Research Lab GNDEC Ludhiana, material Characterization Lab, NITTTR Chandigarh and Lovely Professional University for continuous help and support in the present research.

REFERENCES

Adhikary, K. B., Pang, S., & Staiger, M. P. (2008). Dimensional stability and mechanical behaviour of wood–plastic composites based on recycled and virgin high-density polyethylene (HDPE). *Composites Part B: Engineering*, *39*, 807–815.

Al-Mulla, A., & Gupta, R. K. (2018). Glass-fiber reinforcement as a means of recycling polymers from post-consumer applications. *Journal of Polymers and the Environment*, *26*, 191–199.

Al-Salem, S. M., Lettieri, P., & Baeyens, J. (2009). Recycling and recovery routes of plastic solid waste (PSW): A review. *Waste management*, *29*, 2625–2643.

Al-Salem, S. M., Lettieri, P., & Baeyens, J. (2010). The valorization of plastic solid waste (PSW) by primary to quaternary routes: From re-use to energy and chemicals. *Progress in Energy and Combustion Science*, *36*, 103–129.

Al-Salem, S. M., Antelava, A., Constantinou, A., Manos, G., & Dutta, A. (2017). A review on thermal and catalytic pyrolysis of plastic solid waste (PSW). *Journal of Environmental Management*, *197*, 177–198.

Andersson, T., Stålbom, B., & Wesslén, B. (2004). Degradation of polyethylene during extrusion. II. Degradation of low-density polyethylene, linear low-density polyethylene, and high-density polyethylene in film extrusion. *Journal of Applied Polymer Science*, *91*, 1525–1537.

Anderson, I. (2017). Mechanical properties of specimens 3D printed with virgin and recycled polylactic acid. *3D Printing and Additive Manufacturing*, *4*, 110–115.

Baechler, C., DeVuono, M., & Pearce, J. M. (2013). Distributed recycling of waste polymer into RepRap feedstock. *Rapid Prototyping Journal*, *19*.

Barnes, D. K., Galgani, F., Thompson, R. C., & Barlaz, M. (2009). Accumulation and fragmentation of plastic debris in global environments. *Philosophical transactions of the royal society B: biological sciences*, *364*, 1985–1998.

Brouwer, M. T., & Thoden van Velzen, E. U., Augustinus. A., Soethoudt, H., DeMeester, S., Ragaert, K. (2018). 62–85.

Brüster, B., Addiego, F., Hassouna, F., Ruch, D., Raquez, J. M., & Dubois, P. (2016). Thermo-mechanical degradation of plasticized poly (lactide) after multiple reprocessing to simulate recycling: Multi-scale analysis and underlying mechanisms. *Polymer Degradation and Stability*, *131*, 132–144.

Butler, E., Devlin, G., & McDonnell, K. (2011). Waste polyolefins to liquid fuels via pyrolysis: review of commercial state-of-the-art and recent laboratory research. *Waste and Biomass Valorization*, 2, 227–255.

Çantı, E. (2016). Production and characterization of polymer nanocomposite filaments for 3D printers, Master Thesis. [Production and characterization of polymer nanocomposite filaments for 3d printers] [thesis in Turkish], Dumlupınar University, Kütahya.

Chawla, K., Singh, J., & Singh, R. (2020). On recyclability of thermosetting polymer and wood dust as reinforcement in secondary recycled ABS for nonstructural engineering applications. *Journal of Thermoplastic Composite Materials*, 0892705720925135.

Chawla, K., Singh, R., & Singh, J. (2020). Segregation and Recycling of Plastic Solid Waste: A Review. *Advances in Materials Science and Engineering*, 205–221.

Chong, S., Pan, G. T., Khalid, M., Yang, T. C. K., Hung, S. T., & Huang, C. M. (2017). Physical characterization and pre-assessment of recycled high-density polyethylene as 3D printing material. *Journal of Polymers and the Environment*, *25*, 136–145.

Cisneros-López, E. O., Pal, A. K., Rodriguez, A. U., Wu, F., Misra, M., Mielewski, D. F., & Mohanty, A. K. (2020). Recycled poly (lactic acid)–based 3D printed sustainable biocomposites: a comparative study with injection molding. *Materials Today Sustainability*, *7*, 100027.

Cole, M., Lindeque, P., Halsband, C., & Galloway, T. S. (2011). Microplastics as contaminants in the marine environment: a review. *Marine Pollution Bulletin, 62*, 2588–2597.

Cruz, S. A., Oliveira, É. C., Oliveira, F., Garcia, P. S., & Kaneko, M. L. (2011). Polímeros reciclados para contato com alimentos. *Polímeros, 21*, 340–345.

da Silva, D. J. (2016). Novel preparation technique for polycarbonate/titanium dioxide nanocomposites. *SPE Plast Res Online*, 1–4.

da Silva, D. J., Escote, M. T., Cruz, S. A., Simião, D. F., Zenatti, A., & Curvello, M. S. (2018). Polycarbonate/TiO2 nanofibers nanocomposite: preparation and properties. *Polymer Composites, 39*, E780–E790.

da Silva, D. J., & Wiebeck, H. (2020). Current options for characterizing, sorting, and recycling polymeric waste. *Progress in Rubber, Plastics and Recycling Technology, 36*, 284–303.

D. Gregor-Svetec, M. Leskovšek, U. Vrabič Brodnjak, U. Stankovič Elesini, D. Muck, and R. Urbas (2020). Characteristics of HDPE/cardboard dust 3D printable composite filaments, *Journal of Materials Processing Technology, 276*, 116379, 2020.

Dickson, A. N., Barry, J. N., McDonnell, K. A., & Dowling, D. P. (2017). Fabrication of continuous carbon, glass and Kevlar fibre reinforced polymer composites using additive manufacturing. *Additive Manufacturing, 16*, 146–152.

Dodiuk, H. (Ed.). (2013). *Handbook of thermoset plastics*. William Andrew.

Dodiuk H. and Goodman S. H. (2014). *Handbook of thermoset plastics*. 3rd ed. Oxford: Elsevier.

Duty, C. E., Drye, T., & Franc, A. (2015). *Material development for tooling applications using big area additive manufacturing (BAAM) (No. ORNL/TM-2015/78). Oak Ridge National Lab.(ORNL), Oak Ridge, TN (United States). Manufacturing Demonstration Facility (MDF).*

Duval, C. (2014). Plastic waste and the environment. *Environmental Impact of Polymers*, 13–25.

El Mehdi, M., Benzeid, H., Rodrigue, D., El Kacem, Q., & Bouhfid, R. (2017). Recent advances in polymer recycling: a short review. *Current Organic Synthesis, 14*, 171–185.

Eriksen, M. K., Pivnenko, K., Olsson, M. E., & Astrup, T. F. (2018). Contamination in plastic recycling: Influence of metals on the quality of reprocessed plastic. *Waste Management, 79*, 595–606.

Exconde, M. K. J. E., Co, J. A. A., Manapat, J. Z., & Magdaluyo Jr, E. R. (2019). Materials selection of 3D printing filament and utilization of recycled polyethylene terephthalate (PET) in a redesigned breadboard. *Procedia CIRP, 84*, 28–32.

Farina, I., Fabbrocino, F., Carpentieri, G., Modano, M., Amendola, A., Goodall, R., & Fraternali, F. (2016). On the reinforcement of cement mortars through 3D printed polymeric and metallic fibers. *Composites Part B: Engineering, 90*, 76–85.

Ferreira, R. T. L., Amatte, I. C., Dutra, T. A., & Bürger, D. (2017). Experimental characterization and micrography of 3D printed PLA and PLA reinforced with short carbon fibers. *Composites Part B: Engineering, 124*, 88–100.

Gao, X., Zhang, D., Qi, S., Wen, X., & Su, Y. (2019). Mechanical properties of 3D parts fabricated by fused deposition modeling: Effect of various fillers in polylactide. *Journal of Applied Polymer Science, 136*, 47824.

Gardner, J. M., Sauti, G., Kim, J. W., Cano, R. J., Wincheski, R. A., Stelter, C. J., & Park, S. S. (2016). Additive manufacturing of multifunctional components using high density carbon nanotube yarn filaments. *NF1676L-23685, Nasa Technical Reports Server.*

Garcia, J. M., & Robertson, M. L. (2017). The future of plastics recycling. *Science, 358*, 870–872.

Geyer, R., Jambeck, J. R., & Law, K. L. (2017). Production, use, and fate of all plastics ever made. *Science Advances*, *3*, 1700782.

Gkartzou, E., Koumoulos, E. P., & Charitidis, C. A. (2017). Production and 3D printing processing of bio-based thermoplastic filament. *Manufacturing Review*, *4*, 1.

Hamad, K., Kaseem, M., & Deri, F. (2013). Recycling of waste from polymer materials: An overview of the recent works. *Polymer Degradation and Stability*, *98*, 2801–2812.

Hart, K. R., Frketic, J. B., & Brown, J. R. (2018). Recycling meal-ready-to-eat (MRE) pouches into polymer filament for material extrusion additive manufacturing. *Additive Manufacturing*, *21*, 536–543.

Hill, C., Rowe, K., Bedsole, R., Earle, J., & Kunc, V. (2016). Materials and process development for direct digital manufacturing of vehicles. In SAMPE Long Beach, Conference and Exhibition.

Hodzic, D., & Pandzic, A. (2019). Influence of carbon fibers on mechanical properties of materials in FDM technology. In Proceedings of the 30th DAAAM International Symposium on Intelligent Manufacturing and Automation, Zadar, 1726–9679.

Hwang, S., Reyes, E. I., Moon, K. Ş., Rumpf, R. C., & Kim, N. S. (2015). Thermomechanical characterization of metal/polymer composite filaments and printing parameter study for fused deposition modeling in the 3D printing process. *Journal of Electronic Materials*, *44*, 771–777.

Idrees, M., Jeelani, S., & Rangari, V. (2018). Three-dimensional-printed sustainable biochar-recycled PET composites. *ACS Sustainable Chemistry & Engineering*, *6*, 13940–13948.

Ignatyev, I. A., Thielemans, W., & Vander Beke, B. (2014). Recycling of polymers: a review. *ChemSusChem*, *7*, 1579–1593.

Jambeck, J., Hardesty, B. D., Brooks, A. L., Friend, T., Teleki, K., Fabres, J., ... & Wilcox, C. (2018). Challenges and emerging solutions to the land-based plastic waste issue in Africa. *Marine Policy*, *96*, 256–263.

Jen, Y. M., & Huang, C. Y. (2014). Effect of temperature on fatigue strength of carbon nanotube/epoxy composites. *Journal of Composite Materials*, *48*, 3469–3483.

Karaman, E. and Çolak, O. (2020). The effects of process parameters on mechanical properties and microstructures of parts in fused deposition modeling, *Duzce University Journal of Science & Technology*, *8*, 617–630.

Kozderka, M., Rose, B., Kočí, V., Caillaud, E., & Bahlouli, N. (2016). High impact polypropylene (HIPP) recycling–Mechanical resistance and Lifecycle Assessment (LCA) case study with improved efficiency by preliminary sensitivity analysis. *Journal of Cleaner Production*, *137*, 1004–1017.

Kumar, S., Panda, A. K., & Singh, R. K. (2011). A review on tertiary recycling of high-density polyethylene to fuel. *Resources, Conservation and Recycling*, *55*, 893–910.

Kunc, V. (2015). Advances and challenges in large scale polymer additive manufacturing. In Proceedings of the 15th SPE Automotive Composites Conference, Novi, MI, USA, *9*.

Lanzotti, A., Martorelli, M., Maietta, S., Gerbino, S., Penta, F., & Gloria, A. (2019). A comparison between mechanical properties of specimens 3D printed with virgin and recycled PLA. *Procedia Cirp*, *79*, 143–146.

Li, N., Li, Y., & Liu, S. (2016). Rapid prototyping of continuous carbon fiber reinforced polylactic acid composites by 3D printing. *Journal of Materials Processing Technology*, *238*, 218–225.

Liu, Z., Lei, Q., & Xing, S. (2019). Mechanical characteristics of wood, ceramic, metal and carbon fiber-based PLA composites fabricated by FDM. *Journal of Materials Research and Technology*, *8*, 3741–3751.

Lu, N., & Oza, S. (2013). A comparative study of the mechanical properties of hemp fiber with virgin and recycled high density polyethylene matrix. *Composites Part B: Engineering*, *45*, 1651–1656.

Melenka, G. W., Cheung, B. K., Schofield, J. S., Dawson, M. R., & Carey, J. P. (2016). Evaluation and prediction of the tensile properties of continuous fiber-reinforced 3D printed structures. *Composite Structures*, *153*, 866–875.

Mikula, K., Skrzypczak, D., Izydorczyk, G., Warchoł, J., Moustakas, K., Chojnacka, K., & Witek-Krowiak, A. (2020). 3D printing filament as a second life of waste plastics—a review. *Environmental Science and Pollution Research*, *28*, 12321–12333.

Milios, L., Christensen, L. H., McKinnon, D., Christensen, C., Rasch, M. K., & Eriksen, M. H. (2018). Plastic recycling in the Nordics: A value chain market analysis. *Waste Management*, *76*, 180–189.

Mohammed, M., Das, A., Gomez-Kervin, E., Wilson, D., & Gibson, I. (2017). *EcoPrinting: investigating the use of 100% recycled acrylonitrile butadiene styrene (ABS) for additive manufacturing*.

Mori, K. I., Maeno, T., & Nakagawa, Y. (2014). Dieless forming of carbon fibre reinforced plastic parts using 3D printer. *Procedia Engineering*, *81*, 1595–1600.

Mwanza, B. G., & Mbohwa, C. (2017). Drivers to sustainable plastic solid waste recycling: a review. *Procedia Manufacturing*, *8*, 649–656.

Nabipour, M., Akhoundi, B., & Bagheri Saed, A. (2020). Manufacturing of polymer/metal composites by fused deposition modeling process with polyethylene. *Journal of Applied Polymer Science*, *137*, 48717.

Nagendra, J., & Prasad, M. G. (2020). FDM process parameter optimization by Taguchi technique for augmenting the mechanical properties of nylon–aramid composite used as filament material. *Journal of The Institution of Engineers (India): Series C*, *101*, 313–322.

DeNardo, N. M. (2016). Additive manufacturing of carbon fiber reinforced thermoplastic composites, A Thesis, Purdue University, Indiana.

Ning, F., Cong, W., Qiu, J., Wei, J., & Wang, S. (2015). Additive manufacturing of carbon fiber reinforced thermoplastic composites using fused deposition modeling. *Composites Part B: Engineering*, *80*, 369–378.

Oliveux, G., Dandy, L. O., & Leeke, G. A. (2015). Current status of recycling of fibre reinforced polymers: Review of technologies, reuse and resulting properties. *Progress in Materials Science*, *72*, 61–99.

Onwudili, J. A., Miskolczi, N., Nagy, T., & Lipóczi, G. (2016). Recovery of glass fibre and carbon fibres from reinforced thermosets by batch pyrolysis and investigation of fibre re-using as reinforcement in LDPE matrix. *Composites Part B: Engineering*, *91*, 154–161.

Pan, G. T., Chong, S., Tsai, H. J., Lu, W. H., & Yang, T. C. K. (2018). The effects of iron, silicon, chromium, and aluminum additions on the physical and mechanical properties of recycled 3D printing filaments. *Advances in Polymer Technology*, *37*, 1176–1184.

Patan, Z. (2019). Karbon fiber takviyeli ABS kompozitlerin FDM 3B yazıcı ile üretimi ve "ansys ile modellenmesi", (Doctoral dissertation, Onsekizmart Üniversitesi).

Perez, A. R. T., Roberson, D. A., & Wicker, R. B. (2014). Fracture surface analysis of 3D-printed tensile specimens of novel ABS-based materials. *Journal of Failure Analysis and Prevention*, *14*, 343–353.

Rahimi, A., & García, J. M. (2017). Chemical recycling of waste plastics for new materials production. *Nature Reviews Chemistry*, *1*, 1–11.

Rajpurohit, S. R., & Dave, H. K. (2019). Fused deposition modeling using graphene/pla nano-composite filament. *International Journal of Modern Manufacturing Technologies*, *6*, 2067–3604.

Rezaei, F., Rownaghi, A. A., Monjezi, S., Lively, R. P., & Jones, C. W. (2015). SOx/NOx removal from flue gas streams by solid adsorbents: a review of current challenges and future directions. *Energy & Fuels*, *29*, 5467–5486.

Sahajwalla, V., & Gaikwad, V. (2018). The present and future of e-waste plastics recycling. Current Opinion in Green and Sustainable *Chemistry*, *13*, 102–107.

Sanchez, F. A. C., Lanza, S., Boudaoud, H., Hoppe, S., & Camargo, M. (2015). Polymer Recycling and Additive Manufacturing in an Open Source context: Optimization of processes and methods. In Annual international solid freeform fabrication symposium, (1591–1600).

Shah, J., Snider, B., Clarke, T., Kozutsky, S., Lacki, M., & Hosseini, A. (2019). Large-scale 3D printers for additive manufacturing: design considerations and challenges. *The International Journal of Advanced Manufacturing Technology*, *104*, 3679–3693.

Shent, H., Pugh, R. J., & Forssberg, E. (1999). A review of plastics waste recycling and the flotation of plastics. *Resources, Conservation and Recycling*, *25*, 85–109.

Silvarrey, L. D., & Phan, A. N. (2016). Kinetic study of municipal plastic waste. *International Journal of Hydrogen Energy*, *41*, 16352–16364.

Silveira, A. V. M., Cella, M., Tanabe, E. H., & Bertuol, D. A. (2018). Application of tribo-electrostatic separation in the recycling of plastic wastes. *Process Safety and Environmental Protection*, *114*, 219–228.

Simón, D., Borreguero, A. M., De Lucas, A., & Rodríguez, J. F. (2018). Recycling of polyurethanes from laboratory to industry, a journey towards the sustainability. *Waste Management*, *76*, 147–171.

Singh, R., & Singh, S. (2015). Experimental investigations for statistically controlled solution of FDM assisted Nylon6-Al-Al2O3replica based investment casting. *Materials Today: Proceedings*, *2*, 1876–1885.

Singh, R., Singh, S., & Fraternali, F. (2016a). Development of in-house composite wire based feed stock filaments of fused deposition modelling for wear-resistant materials and structures. *Composites Part B: Engineering*, *98*, 244–249.

Singh, R., Kumar, R., Feo, L., & Fraternali, F. (2016b). Friction welding of dissimilar plastic/polymer materials with metal powder reinforcement for engineering applications. *Composites Part B: Engineering*, *101*, 77–86.

Singh, N., Hui, D., Singh, R., Ahuja, I. P. S., Feo, L., & Fraternali, F. (2017). Recycling of plastic solid waste: A state of art review and future applications. *Composites Part B: Engineering*, *115*, 409–422.

Singh R., Singh H., Farina I. (2019). On the additive manufacturing of an energy storage device from recycled material. *Compos Part B Eng 156*(2019b), 259–265. 10.1016/j.compositesb.2018.08.080.

Spinacé, M. A. D. S., & De Paoli, M. A. (2005). A tecnologia da reciclagem de polímeros. *Química nova*, *28*, 65–72.

Sun, Y., Zwolińska, E., & Chmielewski, A. G. (2016). Abatement technologies for high concentrations of NOx and SO2 removal from exhaust gases: A review. *Critical Reviews in Environmental Science and Technology*, *46*, 119–142.

Tian, X., Liu, T., Wang, Q., Dilmurat, A., Li, D., & Ziegmann, G. (2016a). Recycling and remanufacturing of 3D printed continuous carbon fiber reinforced PLA composites. *Journal of Cleaner Production*, *30*, 1e10.

Tian, X., Liu, T., Yang, C., Wang, Q., & Li, D. (2016b). Interface and performance of 3D printed continuous carbon fiber reinforced PLA composites. *Composites Part A: Applied Science and Manufacturing*, *88*, 198–205.

Unnisa, S. A., & Hassanpour, M. (2017). Development circumstances of four recycling industries (used motor oil, acidic sludge, plastic wastes and blown bitumen) in the world. *Renewable and Sustainable Energy Reviews*, *72*, 605–624.

Van Der Klift, F., Koga, Y., Todoroki, A., Ueda, M., Hirano, Y., & Matsuzaki, R. (2016). 3D printing of continuous carbon fibre reinforced thermo-plastic (CFRTP) tensile test specimens. *Open Journal of Composite Materials*, *6*, 18.

Vickers, N. J. (2017). Animal communication: when i'm calling you, will you answer too. *Current Biology*, *27*, 713–R715.

Woern, A. L., Byard, D. J., Oakley, R. B., Fiedler, M. J., Snabes, S. L., & Pearce, J. M. (2018). *Fused particle fabrication 3-D printing*.

Yildirir, E., Miskolczi, N., Onwudili, J. A., Németh, K. E., Williams, P. T., & Sója, J. (2015). Evaluating the mechanical properties of reinforced LDPE composites made with carbon fibres recovered via solvothermal processing. *Composites Part B: Engineering*, *78*, 393–400.

Zander, N. E., Gillan, M., & Lambeth, R. H. (2018). Recycled polyethylene terephthalate as a new FFF feedstock material. *Additive Manufacturing*, *21*, 174–182.

Zander, N. E., Gillan, M., Burckhard, Z., & Gardea, F. (2019). Recycled polypropylene blends as novel 3D printing materials. *Additive Manufacturing*, *25*, 122–130.

Zhao P., Rao C., Gu F., Sharmin N., Fu J. (2018a). Close-looped recycling of polylactic acid used in 3D printing: An experimental investigation and life cycle assessment. *Journal of Cleaner Production, 197*: 1046–1055, 10.1016/j.jclepro.2018.06.275

Zhao X. G., Hwang K. J., Lee D., Kim T., Kim N. (2018b). Enhanced mechanical properties of self-polymerized polydopamine-coated recycled PLA filament used in 3D printing. *Applied Surface Science, 441* (2018b), 381–387.

Zhong, W., Li, F., Zhang, Z., Song, L., & Li, Z. (2001). Short fiber reinforced composites for fused deposition modeling. *Materials Science and Engineering: A*, *301*, 125–130.

6 Tertiary Recycling of Plastic Solid Waste for Additive Manufacturing

Vinay Kumar, Rupinder Singh, and Inderpreet Singh Ahuja

6.1 INTRODUCTION

In the past two decades, thermoplastics have been investigated extensively by many researchers for numerous engineering and biomedical applications. Such investigations outlined cost-effective technologies for processing polymeric materials like polyamide-6 (PA6), acrylonitrile butadiene styrene (ABS), polylactic acid (PLA), PVDF, etc. to utilize such polymers in manufacturing practices through friction spot welding, friction stir welding, and the AM process (Kumar et al., 2020a). Studies reported on reinforcement of natural and synthetic fibres in polymers have been explored for preparing the polymer matrix composite (PMC) for advanced manufacturing processes and fabricate eco-friendly composite products to reduce the environmental impact (Kumar et al., 2020b). Similarly, nanoparticles as reinforcement for PMC also have been investigated for AM and rapid prototyping. It has been observed that mechanical strength, wear, and morphological properties of a base polymer matrix increased after the addition of such nanoparticles (Begum et al., 2020). It has been outlined that acceptable rheological characteristics in terms of MFI is required to be established in PMCs so that such compositions/proportions can be easily used for useful 3D-printing applications. Acceptable rheological property has been investigated of PMC that ensures the maximum possible reinforcement of nanoparticles in the specific thermoplastic grade. The non-conventional material composites prepared by such reinforcements have reportedly large engineering applications (Chohan et al., 2020, Boparai & Singh, 2021).

Some reinforcements like G_{NP}, carbon nanoparticles (C_{NP}), and $BaTiO_3$ have been used for preparing a composite matrix of PVDF by secondary (2°) and 3° recycling processes for developing piezoelectric sensors for electrical equipments that can be 3D printed to manufacture the products on a large scale (Boparai et al., 2020, Kumar et al., 2021a, Kumar et al., 2020b). Based upon reported studies, a detailed literature survey was performed to explore the various recycling processes for handling PSW for different engineering practices. Table 6.1 shows the list of various key terms surveyed in literature review as per the Web of Science database.

DOI: 10.1201/9781003184164-6

TABLE 6.1
List of key terms investigated for recycling PSW for various applications (as per the Web of Science database)

ID	Term	Occurrences	Relevance Score
1	Activity performed on recycled polymers	3	0.6884
2	Additive manufacturing	2	3.3406
3	Adsorption spectrum of polymers	2	0.9981
4	Advantage of polymer recycling	2	1.108
5	Art review for reusing polymers	3	2.072
6	Average life cycle of thermoplastics	3	0.4839
7	Biological recycling	2	2.0803
8	Chemical reactivity of plastics	2	0.9709
9	Catalyst for boosting polymer blending	11	0.6465
10	Catalytic degradation	2	0.6551
11	Cement paste for plastic tiles	2	1.2553
12	Change in properties after processing	3	0.4681
13	Chemical oxygen demand	3	0.759
14	Chemical recycling method	2	2.2594
15	Chlorination application	2	0.8007
16	Circular economy	4	1.0694
17	Chemical-assisted mechanical blending	2	0.8588
18	Construction using recycled plastic	2	0.865
19	Collection challenges for waste plastics	3	0.7197
20	Component manufacturing	2	0.874
21	Concrete blended with polymers	2	1.2926
22	Construction applications	4	1.0669
23	Content for improving polymers performance	7	0.3528
24	Contribution for waste management	3	0.3505
25	Conversion to useful composites	7	0.7538
26	Contribution of organizations	3	0.3981
27	Cross-linking of consumer and manufacturer	2	2.0259
28	Decay with time	2	0.751
29	Degradation of ecosystem	4	0.7935
30	Degrees of recycling	3	0.6881
31	Demolition waste	2	1.2553
32	Economic issue	2	1.9729
33	Effluent containing plastic waste	6	0.7262
34	Energy recovery for circular economy	3	1.0802
35	*Escherichia coli* for polymer recycling	2	1.3943
36	Ethyl benzene reactivity	2	1.0054
37	Example for 4 R policy	4	1.3337
38	Experimental data for recycled waste	5	0.507
39	Factor affecting reuse	2	1.0666
40	Filament fabrication	2	0.865

TABLE 6.1 (Continued)
List of key terms investigated for recycling PSW for various applications (as per the Web of Science database)

ID	Term	Occurrences	Relevance Score
41	Feedstock recycling	2	1.1229
42	Feldspar	3	1.0514
43	Filter material from plastics	2	1.0257
44	Focus on economy	3	1.1485
45	Form of used polymers	2	1.6447
46	Formation of composite matrices	3	0.6823
47	Fuels for future generations	5	0.6021
48	Gasoline applications	2	0.865
49	Good yield polymer applications	2	0.9526
50	Guideline for using polymer grades	4	0.7672
51	HDPE AM applications	2	1.3455
52	High-impact polystyrene 3D printing	2	1.3455
53	High purity grade polymers	2	0.6553
54	Hospital products from AM	3	1.1223
55	Hour-wise consumption in industries	2	1.0388
56	Hydrocarbon arrangement	3	0.8034
57	Hydrogen content	2	1.0051
58	Implication of plastic material	2	1.1669
59	Incidence for reduced polymer applications	2	1.1665
60	Indicator of environmental issues	2	1.016
61	Influence of PSW	3	0.8523
62	Liquid product packaging	3	0.7388
63	Literature available for polymer recycling	2	0.6612
64	Location of recycling setups	2	0.8899
65	Loop of material recycling	2	2.2594
66	Loss of material quality	3	0.4808
67	Low-density polyethylene applications	3	0.7152
68	Maximum mechanical strength	2	0.6305
69	Mechanical recycling	3	0.7414
70	Membrane fabrication	2	1.1547
71	Mineral processing	2	1.2553
72	Mixture of polymers	3	0.6651
73	Natural aggregate	2	1.2926
74	Natural resource	2	0.9826
75	Need of biomedical implants	2	0.635
76	Particle reinforcements	3	1.2101
77	Phosphorus nanoparticles	3	0.7152
78	Pilot studies on recycling methods	2	0.7972
79	Plastic material	2	0.9946

(Continued)

TABLE 6.1 (Continued)
List of key terms investigated for recycling PSW for various applications (as per the Web of Science database)

ID	Term	Occurrences	Relevance Score
80	Plastic recycling	2	1.5068
81	Polyethylene applications	2	0.5883
82	Polymer degradation	2	0.5944
83	Polymer waste	3	0.8497
84	Polystyrene for AM	2	1.0054
85	Processing of polymer composites	3	1.202
86	Product slate	2	0.6718
87	Progress in composite applications	3	1.4907
88	Promising candidate for 3D printing	2	1.9729
89	Promising technology for advance manufacturing	2	1.376
90	Quartz recycling	2	0.8871
91	Range of recyclable polymers	9	0.268
92	Reaction temperature for smart composites	2	0.8715
93	Reaction time study	3	0.8114
94	Recycled glass for industrial use	2	1.0257
95	Recycled sand applications	2	1.2553
96	Recycled water	4	0.9702
97	Reinforcement in polymer recycling	2	1.9394
98	Relation in base matrix and reinforcements	2	0.7853
99	Removal of surface defects	7	0.4453
100	Risk in recycling methods	3	1.1862
101	Sand as a reinforcement	5	0.5089
102	Sand medium for plastic landfills	2	1.0257
103	Solid waste management	2	0.7972
104	Stages of recycling	3	0.7698
105	Standard operation	2	0.865
106	Styrene monomer for 3D printing	2	0.7943
107	Suitability of recycled polymers	2	0.2452
108	Tertiary amine applications	2	0.9106
109	Tertiary care steps	2	1.1665
110	Tertiary filter medium	2	1.0257
111	Tertiary treatment of polymers	4	0.8414
112	Thermoplastic and thermosetting recycling	2	2.4272
113	Toluene in AM	3	0.8357
114	Total energy consumption	3	1.0017
115	Transportation cost to recycling units	2	0.7510
116	Treatment of waste plastics	10	0.4979
117	Twin-screw extrusion of polymers	2	1.0257
118	Valuable product from recycling methods	2	0.9131
119	Wastewater treatment	4	0.8488

TABLE 6.1 (Continued)
List of key terms investigated for recycling PSW for various applications (as per the Web of Science database)

ID	Term	Occurrences	Relevance Score
120	Waste plastic treatment plant	2	0.7429
121	Solid plastic waste recycling	2	1.1547
122	World with/without recycled products	3	0.7299
123	Yield strength of recycled polymers	10	0.5667

The literature reported in Table 6.1 on various key terms related to recycling of thermoplastic and thermosetting for 3D printing and other industrial applications has highlighted that smart 4D-capable composites and polymer ceramic composites may be used effectively for providing tunable or customizable functional prototypes as 3D-printed solutions to solve repair and maintenance-related problems of heritage structures (Sharma et al., 2020, Sharma et al., 2021). The keywords reported in Table 6.1 were used to obtain a web/cluster of research terms to illustrate the interconnectivity of researchers for resolving environmental issues caused by polymers. Figure 6.1 shows the web of key research terms investigated for improving application areas of polymers after specific recycling techniques.

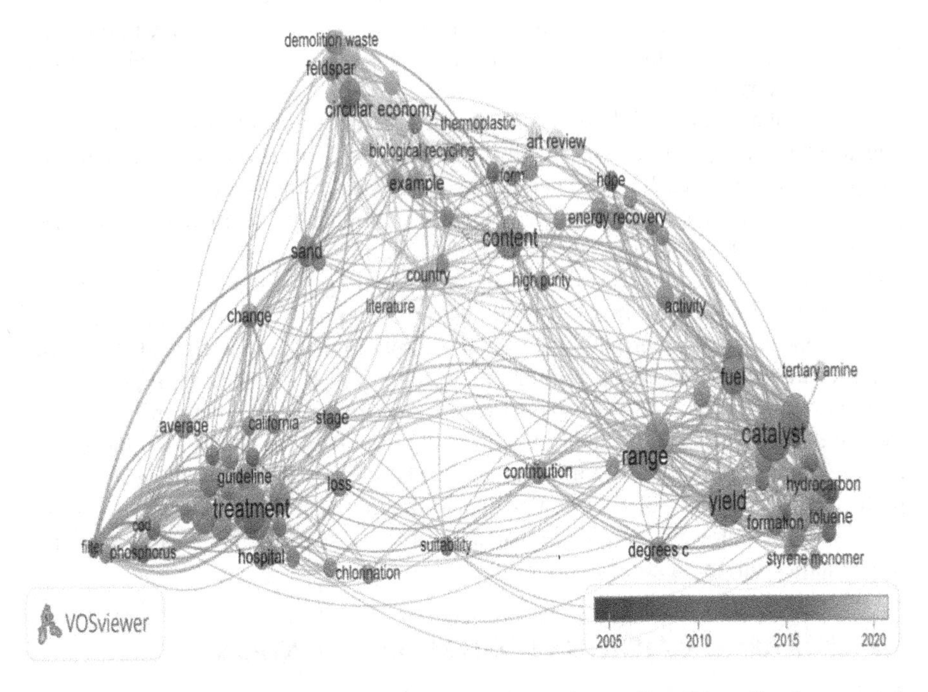

FIGURE 6.1 Web of research terms investigated for improving the application area of polymers.

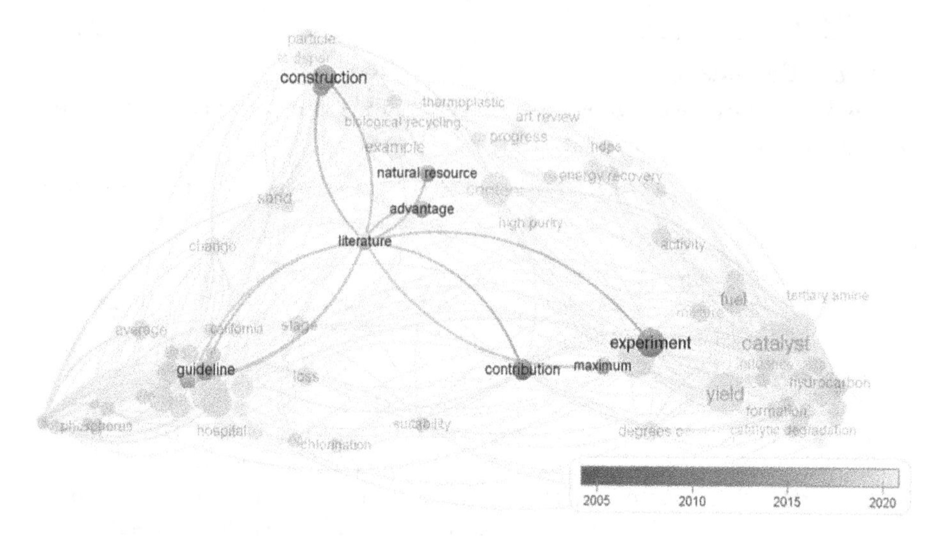

FIGURE 6.2 Key terms highlighting the application of recycled materials in construction.

The studies reported on secondary recycling of PA6, PLA, and PVDF polymers using hybrid reinforcements have indicated that such composites matrices may be used for manufacturing sports products (like shoes), biomedical implants for veterinary patients, and 4D-capable optical sensors and equipments (Kumar et al., 2020c, Kumar et al., 2020d, Singh and Singh 2017, Kumar et al., 2021c). Figure 6.2 shows the use of recycled materials for construction activities so that the burden can be reduced from natural materials and more advantages may be obtained from recycled materials.

ABS is one of the widely used engineering plastic especially for AM processes due to which recycled ABS is of great demand for fabrication of functional prototypes. Some studies have revealed that cryogenic milling and screw extrusion processing of primary (1°) recycled ABS may contribute to increasing the mechanical strength of ABS (Singh et al., 2021, Kumar et al., 2020e, Kumar et al., 2021d). The studies reported on mechanical blending and chemical-assisted mechanical blending of polymers for composite development has outlined that rheological, mechanical, thermal, wear, and morphological properties of the base polymer matrix increased on reinforcement with G_{NP} that not only improved the lubrication and flow-ability of the ABS but also imparted magnetic properties in the base matrix of ABS (Kumar et al., 2020f, Kumar et al., 2020g, Kumar et al., 2021e, Kumar et al., 2021f, Kumar et al., 2021g). The experimental investigations performed on various polymer composites outlined that recycling of waste plastics to prepare a composite may be considered as a novel route to manage PSW for useful industrial applications.

6.2 RESEARCH GAP AND PROBLEM FORMULATION

The literature survey shows that significant work has been reported for highlighting the suitable and effective material processing and recycling routes to manage solid

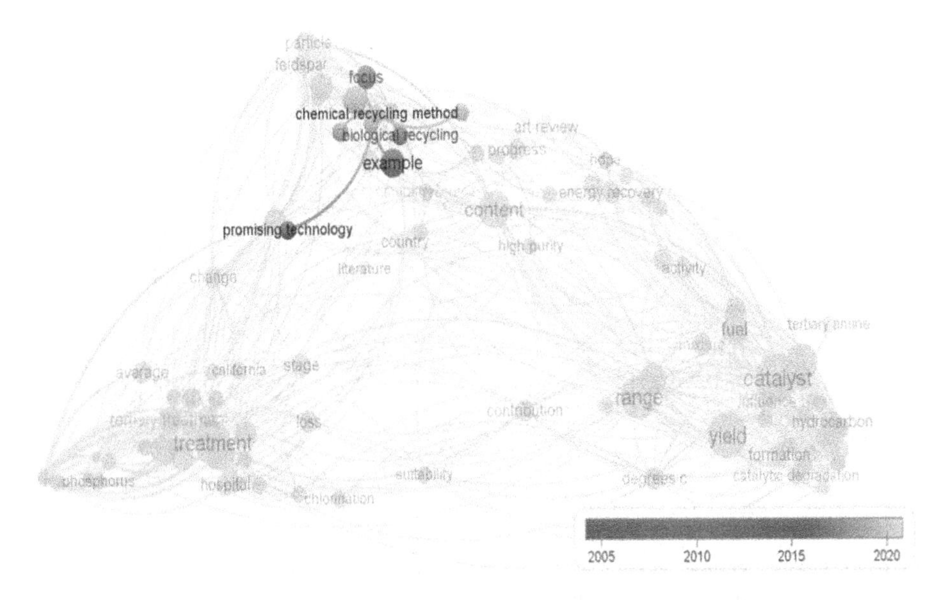

FIGURE 6.3 Literature gap in web of key terms for chemical recycling methods and 3D printing.

waste of thermoplastics and thermosetting. But hitherto, very little has been reported on 3° recycling as a standard material processing route and process capability analysis of the proposed process to recycle PSW for preparation of polymer matrix-based composite that can be processed in a screw extrusion setup by considering optimized settings for fabrication of feedstock filament for 3D/4D-printing functional prototypes. Figure 6.3 highlights the gap in literature that chemical recycling methods are least reported for processing waste plastic–based composites for 3D-printing applications.

The literature gap also shows that various AM approaches like FDM, stereolithography, etc. are weakly linked to utilize chemically recycled plastics for 3D-printing applications. Figure 6.4 shows that there is no/weak linking of recycling processes for AM. The investigations performed on 4D capabilities of secondary (2°) recycled PVDF by reinforcement of G_{NP} and calcium carbonate ($CaCO_3$) highlighted that a smart composite may be used effectively for 4D applications with customizable and tunable properties.

The present work is an extension of previously reported work in which recycling of PVDF is reported for AM-based 3D/4D applications. 3° recycling of PVDF was performed by blending G_{NP} and MnZnO in the presence of a DMF solvent to prepare the composite for 3D-printing functional-graded specimens (Kumar et al., 2021d,e). The properties tested for the composition/proportion like: MFI, density, viscosity, thermal stability, heat capacity, mechanical strength, and porosity were also analyzed for process capability and performance to establish a standard material processing process for recycled PVDF that may be followed to produce 3D-printed prototypes for various engineering applications.

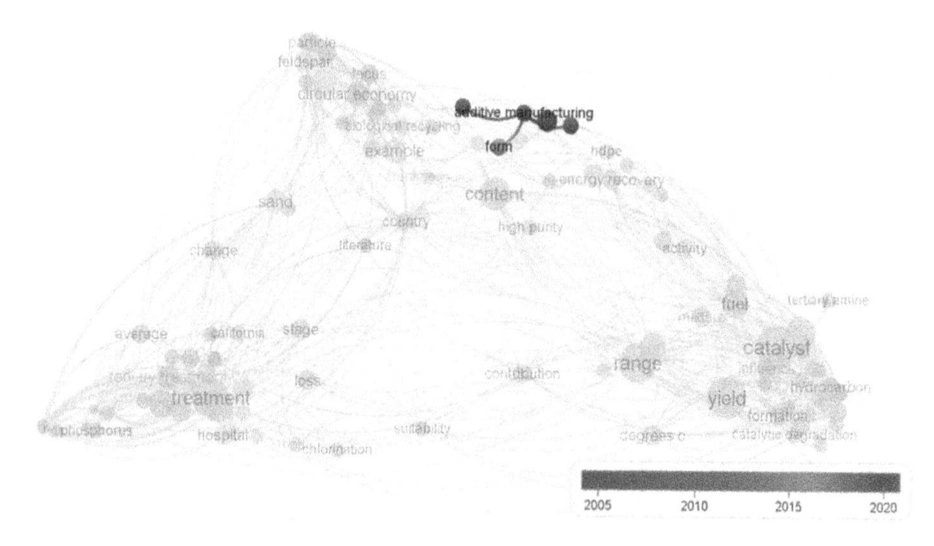

FIGURE 6.4 Literature gap for weak linking of additive manufacturing and its domains with circular economy and chemical recycling methods.

6.3 EXPERIMENTATION

The 3° recycling of waste PVDF was performed by dissolving used PVDF pallets in a DMF solvent at 55°C and G_{NP} and MnZnO were mixed with the solution uniformly with constant stirring. The blends of PVDF-G_{NP}-MnZnO composition/proportion were prepared by mixing the reinforcements in different proportions by weight. On completion of the blending process, the compositions/proportions were dried in a vacuum oven to obtain dry lumps of the composite. These lumps were investigated for rheological analysis including MFI, density, and viscosity of the composite prepared by 3° recycling. The suitable compositions/proportions were then tested for thermal property by differential scanning calorimetry (DSC)–based thermal analysis. On the basis of rheological and thermal properties, the best composition of PVDF-G_{NP}-MnZnO composite was twin-screw (TS) extruded to investigate the mechanical properties of the composite by using the universal testing machine.

The morphological properties i.e. porosity and shore-D hardness (S_DH) of composite samples were observed to investigate the effect of the 3° recycling process on the PVDF composite for 3D-printing applications. Finally, the process capability and performance capability of the proposed recycling method was investigated in terms of process capability (C_p), process capability index (C_{pk}), process performance (P_p), process performance index (P_{pk}), and overall standard deviation for process capability (σ).

6.4 RESULTS AND DISCUSSION

6.4.1 Rheological Analysis

The results obtained for rheological properties of various PVDF-G_{NP}-MnZnO–blended compositions/proportions are listed in Table 6.2. It was observed that

TABLE 6.2
Results obtained for rheological properties: MFI, density, and viscosity of PVDF composite

S. No.	Composition/Proportion	MFI	Density	Viscosity
1	PVDF	3.767	1.014	4032.14
2	PVDF-1G_{NP}-1MnZnO	3.743	1.016	4064.72
3	PVDF-1G_{NP}-2MnZnO	3.689	1.017	4097.17
4	PVDF-1G_{NP}-3MnZnO	3.521	1.021	4121.57
5	PVDF-1G_{NP}-4MnZnO	3.497	1.023	4134.28
6	PVDF-2G_{NP}-1MnZnO	3.471	1.026	4149.79
7	PVDF-2G_{NP}-2MnZnO	3.463	1.039	4156.43
8	PVDF-2G_{NP}-3MnZnO	3.451	1.047	4167.35
9	PVDF-2G_{NP}-4MnZnO	3.444	1.114	4186.76
10	PVDF-3G_{NP}-1MnZnO	3.396	1.187	4194.91
11	PVDF-3G_{NP}-2MnZnO	3.347	1.224	4214.33
12	PVDF-3G_{NP}-3MnZnO	3.311	1.265	4229.52
13	PVDF-3G_{NP}-4MnZnO	3.258	1.287	4251.47
14	PVDF-4G_{NP}-1MnZnO	3.204	1.338	4265.73
15	PVDF-4G_{NP}-2MnZnO	3.185	1.375	4273.66
16	PVDF-4G_{NP}-3MnZnO	3.129	1.384	4279.01
17	PVDF-4G_{NP}-4MnZnO	3.084	1.391	4286.37
18	PVDF-5G_{NP}-1MnZnO	3.011	1.399	4297.92
19	PVDF-5G_{NP}-2MnZnO	2.985	1.403	4321.38
20	PVDF-5G_{NP}-3MnZnO	2.976	1.429	4353.81
21	PVDF-5G_{NP}-4MnZnO	2.87	1.478	4378.27
22	PVDF-6G_{NP}-1MnZnO	2.832	1.501	4391.3
23	PVDF-6G_{NP}-2MnZnO	2.789	1.512	4408.05
24	PVDF-6G_{NP}-3MnZnO	2.711	1.519	4411.34
25	PVDF-6G_{NP}-4MnZnO	2.934	1.624	4396.74
26	PVDF-7G_{NP}-1MnZnO	3.015	1.673	4309.32
27	PVDF-8G_{NP}-1MnZnO	3.147	1.362	4285.9
28	PVDF-8G_{NP}-2MnZnO	3.199	1.224	4251.21
29	PVDF-8G_{NP}-3MnZnO	3.489	1.201	4194.37
30	PVDF-8G_{NP}-4MnZnO	3.592	1.827	4187.22

the MFI of the composite first decreased with an increase in G_{NP} and MnZnO weight proportions due to which density and viscosity of the compositions/proportions increased. But, an increase of G_{NP} and MnZnO reinforcements content up to 8% and 4%, respectively, increased the MFI of the composite and ultimately choked the MFI tester, indicating no possible way for more reinforcement in the base PVDF matrix.

TABLE 6.3
Selected compositions/proportions of PVDF composites and their average rheological properties

S. No.	Composition/ Proportion	Average MFI g/(10 min)	Average Density g/cm^3	Average Viscosity Pa-s
1	PVDF	3.76 ± 0.002	1.017	4031.87 ± 0.03
2	PVDF-2G_{NP}-1MnZnO	3.47 ± 0.004	1.022	4148.91 ± 0.04
3	PVDF-4G_{NP}-2MnZnO	3.18 ± 0.003	1.379	4214.15 ± 0.03
4	PVDF-6G_{NP}-3MnZnO	2.71 ± 0.002	1.512	4411.44 ± 0.02
5	PVDF-8G_{NP}-4MnZnO	3.59 ± 0.003	1.836	4186.99 ± 0.03

Based on the results obtained in Table 6.2, 05 compositions/proportions shown in Table 6.3 were selected in which significant rheological characteristics were obtained for the composite to further investigate the thermal properties of PVDF and its various composite samples. Table 6.3 highlights the average MFI, density, and viscosity of selected compositions/proportions of PVDF-G_{NP}-MnZnO composites.

6.4.2 Thermal Analysis

The DSC test of selected compositions/proportions shows that the PVDF-G_{NP}-MnZnO composites are thermally stable for AM applications as acceptable heat capacity was observed in each tested sample. Table 6.4 shows the results obtained for average thermal heat capacity (J/g) of PVDF and PVDF-G_{NP}-MnZnO composites.

It was observed from rheological and thermal analysis that sample 4 as per Tables 6.3 and Table 6.4 outlined the best acceptable flowability and thermal properties for feedstock filament preparation and 3D-printing functional prototypes. The composition PVDF-6G_{NP}-3MnZnO was finally selected for TS extrusion for 3D-printable filament fabrication. The Taguchi L9 orthogonal array was used to

TABLE 6.4
DSC results for heat capacity of PVDF and PVDF-G_{NP}-MnZnO composites

S. No.	Composition/Proportion	Average Thermal (heat) Capacity (J/g)
1	PVDF	27.14
2	PVDF-2G_{NP}-1MnZnO	29.49
3	PVDF-4G_{NP}-2MnZnO	33.58
4	PVDF-6G_{NP}-3MnZnO	86.41
5	PVDF-8G_{NP}-4MnZnO	34.65

TABLE 6.5
Selected input parameters of TS extruder and their levels

S. No.	Screw Temperature (°C)	Screw Speed (rpm)	Load (kg)
1	195	75	5
2	205	85	8
3	215	95	11

TABLE 6.6
Design of experiment performed on TS setup for PVDF-6G_{NP}-3MnZnO composite

S. No.	Screw Temperature (°C)	Screw Speed (rpm)	Load (kg)
1	195	75	5
2	195	85	8
3	195	95	11
4	205	75	8
5	205	85	11
6	205	95	5
7	215	75	11
8	215	85	5
9	215	95	8

obtain the best settings for filament fabrication. Table 6.5 shows the process parameters selected for TS extrusion of the PVDF-6G_{NP}-3MnZnO composite. Table 6.6 shows the design of experiment (DOE) to prepare wire samples of the composite for testing mechanical properties of the composite when processed at different extrusion conditions.

6.4.3 Mechanical Analysis

The wire samples of the PVDF-6G_{NP}-3MnZnO composite prepared as per Table 6.6 were tested for mechanical properties like peak strength (PS), break strength (BS), Young's modulus etc. on a UTM setup by performing a tensile test to obtain the best extrusion processing conditions. Table 6.7 shows the results obtained for UTM testing of nine PVDF-6G_{NP}-3MnZnO composite wire samples.

It was observed that the twin-screw temperature of 215°C, screw speed of 95 rpm, and load of 5 kg contributed towards the best acceptable mechanical properties in the composite in terms of Young's modulus to impart good elasticity in the material. The PS of 21.73 N/mm^2 was observed in the best wire sample i.e. sample 9 as per Table 6.6.

TABLE 6.7

UTM results obtained for PVDF-6G_{NP}-3MnZnO composite wire samples

S. No.	Peak Load (N)	Peak Elongation (mm)	Break Load (N)	Break Elongation (mm)	Peak Strength (N/mm^2)	Break Strength (N/mm^2)	Peak Strain	Break Strain	Modulus of Toughness	Young's Modulus (MPa)
1	81.1	2.41	73.6	2.43	16.64	15.05	0.045	0.051	0.40	338.3
2	78.2	1.78	70.5	1.81	15.95	14.39	0.032	0.033	0.26	425.4
3	119.3	3.09	107.6	3.21	24.31	21.88	0.064	0.070	0.82	396.2
4	124.1	2.67	112.0	2.52	25.32	22.81	0.056	0.059	0.66	465.9
5	129.3	3.04	116.9	3.09	26.45	23.79	0.061	0.064	0.74	428.9
6	131.4	2.45	118.7	2.51	26.84	24.25	0.051	0.059	0.70	534.2
7	104.1	1.42	94.4	1.45	21.31	19.29	0.033	0.038	0.35	732.2
8	111.2	1.81	100.1	1.88	22.64	20.45	0.034	0.059	0.55	611.8
9	106.3	1.44	96.1	1.66	21.73	19.63	0.031	0.038	0.32	747.6

6.4.4 Process Capability Analysis

For process capability analysis, nine samples of the PVDF-6G_{NP}-3MnZnO composite were prepared at a single setting number 9 as per DOE in Table 6.6. After obtaining the PS of each sample, the S_DH and porosity of tensile-tested samples were performed along the cross-section area. Table 6.8 show the PS, S_DH, and porosity of PVDF-6G_{NP}-3MnZnO composite samples.

The results obtained for PS and S_DH were processed in the Process Capability Wizard software package developed by Symphony Technologies, USA to analyze the effect of extrusion process parameters on mechanical and morphological properties of the composite. The outcomes of the analysis highlighted that process variables may be controlled effectively for fabrication of composite filament for the AM-based FDM process. The achievement of investigated properties in the desired tolerance limits indicated better output in terms of good mechanical properties and standard process capability. The capability statistics in terms of C_p, C_{pk}, P_p, P_{pk}, and σ for 3° recycling of solid PVDF waste composite were obtained for AM applications. The results also highlighted that the experimental observations of PS and S_DH passed the Anderson-Darling (AD) test, indicating that the sample testing range comes under the category of normal distribution. Figure 6.5 shows the histogram and normal distribution plots PS of PVDF-6G_{NP}-3MnZnO composite prepared by the 3° recycling process.

The upper specification limit (USL) and lower specification limit (LSL) for PS was considered 22.5 and 21.0, respectively. With standard deviation, $\sigma = 0.1329$ for potential process capability, $C_p = 1.55$, $C_{pk} = 1.43$, $P_p = 1.33$, and $P_{pk} = 1.37$ was observed for the PS of the PVDF composite. Similarly, the USL = 66.5 and LSL = 60 were selected for S_DH. Figure 6.6 shows the histogram and normal distribution plots S_DH of the PVDF-6G_{NP}-3MnZnO composite prepared by the 3° recycling process.

TABLE 6.8
Peak strength, shore-D hardness and porosity of PVDF-6G_{NP}-3MnZnO samples

S. No.	Peak Strength (N/mm^2)	Shore-D Hardness (HD)	Porosity (%)
1	21.5	61.5	11.3
2	21.6	63.5	10.9
3	21.5	62.4	12.5
4	21.7	62.5	10.8
5	21.5	63.1	11.6
6	21.7	62.8	12.3
7	21.6	63.3	11.5
8	21.5	63.5	12.4
9	21.7	62.5	12.2

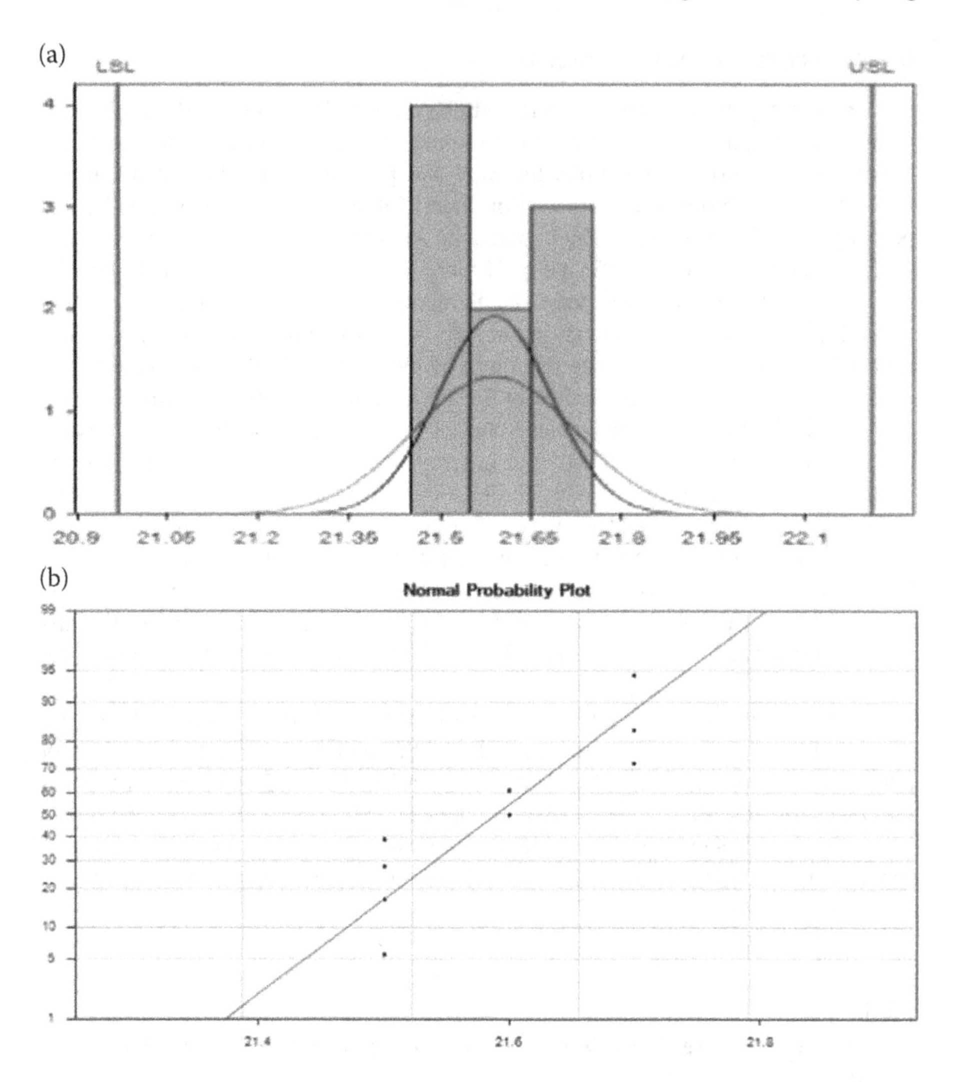

FIGURE 6.5 Histogram (a) and normal distribution (b) plots for PS of PVDF composite.

For the selected USL and LSL, $\sigma = 0.8311$ was observed for potential process capability. On the other hand, $\sigma = 0.8569$ was obtained for overall capability of the process. For less than 5% error in the process with a 95% confidence level, the capability statistic shows that obtained results passed the A-D test for acceptable S_DH in the composite for batch production activity. The process capability indices $C_p = 1.47$, $C_{pk} = 1.42$, $P_p = 1.43$, and $P_{pk} = 1.34$ were obtained for potential capability and overall process capability in regards to the SDH of the PVDF composite prepared by the 3° recycling process.

The analysis outlined that the 3° recycling of the PVDF for fabrication of the 3D-printer feedstock filament resulted in a useful material processing route to fabricate functional prototypes of the proposed composite for 3D/4D-based AM processes.

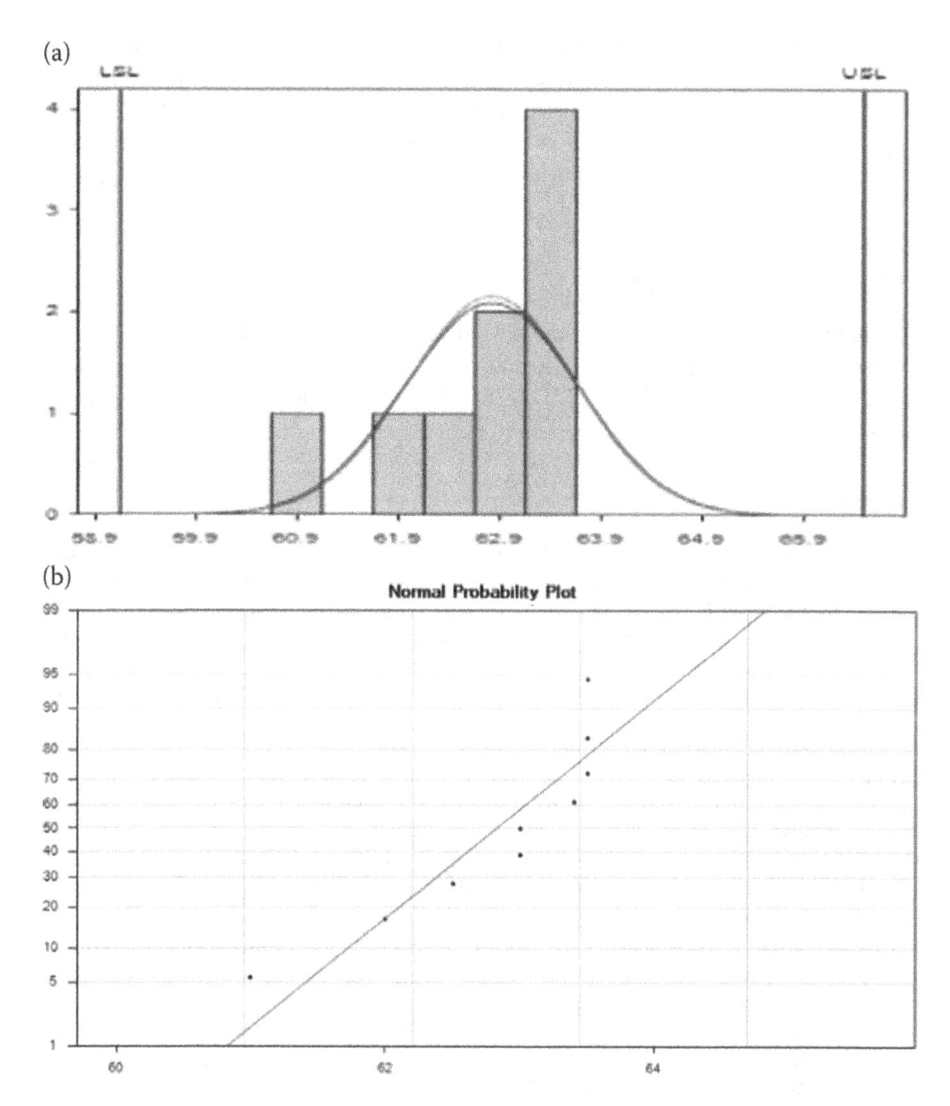

FIGURE 6.6 Histogram (a) and normal distribution (b) plots for SDH of PVDF composite.

6.5 SUMMARY

The 3° recycling of the PVDF PSW blended with G_{NP} and MnZnO at different weight proportions may be considered an acceptable material processing route for FDM-based AM processes like 3D/4D printing. The rheological analysis of compositions/proportions highlighted that MFI of the base polymer matrix decreased in a composite format up to a certain limit and then it increased. The thermal analysis of the selected compositions/proportions outlined that PVDF-6G_{NP}-3MnZnO is the most acceptable composition for AM applications due to its better rheological and thermal stability characteristics. On TS extrusion and UTM testing, it was obtained that the twin-screw temperature of 215°C, screw speed of 95 rpm, and load of 5 kg contributed towards the best acceptable mechanical properties in the composite.

Further, investigations on process capability of composite support the research outcomes that the 3° recycling of solid PVDF waste for composite manufacturing and feedstock filament fabrication is very useful for various engineering applications through AM processes like FDM.

ACKNOWLEDGEMENT

The authors are grateful to the manufacturing research lab, GNDEC, Ludhiana for lab facilities, and Department of Science and Technology (DST)-Government of India (under Project File No DST/TDT/SHRI-35/2018) for providing the financial support in the present work.

REFERENCES

Begum S., Fawzia S., & Hashmi M. S. J. (2020). Polymer matrix composite with natural and synthetic fibres. *Advances in Materials and Processing Technologies*, *6*(3), 547–564.

Boparai K. S., Singh R., & Hashmi M. S. J. (2020). Reinforced non-conventional material composites: A comprehensive review. *Advances in Materials and Processing Technologies*, *7*, 1–10.

Boparai, K. S., & Singh, R., (2021). On decision making of reinforcement proportion in thermoplastic matrix based upon rheological properties: An overview. *Advances in Materials and Processing Technologies*, pp. 1–14.

Chohan J. S., Boparai K. S., Singh R., & Hashmi M. S. J. (2020). Manufacturing techniques and applications of polymer matrix composites: A brief review. *Advances in Materials and Processing Technologies*, 1–11.

Kumar R., Singh R., Ahuja I. P. S., & Hashmi M. S. J. (2020a). Processing techniques of polymeric materials and their reinforced composites. *Advances in Materials and Processing Technologies*, *6*(3), 591–607.

Kumar V., Singh R., & Ahuja I. P. S. (2020b). Effect of extrusion parameters on primary recycled ABS: Mechanical, rheological, morphological and thermal properties. *Materials Research Express*, *7*(1), 015208.

Kumar V., Singh R., Ahuja I. P. S., & Hashmi M. S. J. (2020c). On technological solutions for repair and rehabilitation of heritage sites: A review. *Advances in Materials and Processing Technologies*, *6*(1), 146–166.

Kumar R., Singh R., & Hashmi M. S. J. (2020d). Polymer-Ceramic composites: A state of art review and future applications. *Advances in Materials and Processing Technologies*, 1–14.

Kumar V., Singh R., & Ahuja I. P. S. (2020e). On cryogenic milling of primary recycled ABS: Rheological, morphological, and surface properties. *Journal of Thermoplastic Composite Materials*, 0892705720932621.

Kumar R., Pandey A. K., Singh R., & Kumar V. (2020f). On nano polypyrrole and carbon nano tube reinforced PVDF for 3D printing applications: Rheological, thermal, electrical, mechanical, morphological characterization. *Journal of Composite Materials*, *54*(29), 4677–4689.

Kumar V., Singh R., & Ahuja I. P. S. (2020g). Secondary recycled acrylonitrile–butadiene–styrene and graphene composite for 3D/4D applications: Rheological, thermal, magnetometric, and mechanical analyses. *Journal of Thermoplastic Composite Materials*, 0892705720925114.

Kumar V., Singh R., & Ahuja I. P. S. (2021a). On Correlation of Rheological, Thermal, Mechanical and Morphological Properties of Mechanically Blended PVDF-Graphene Composite for 4d Applications. Reference Module in Materials Science and Materials Engineering, doi.org/10.1016/B978-0-12-820352-1.00192-9.

Kumar V., Singh R., & Ahuja I. P. S. (2021b). On correlation of rheological, thermal, mechanical and morphological properties of chemical assisted mechanically blended ABS-Graphene composite as tertiary recycling for 3D printing applications. *Advances in Materials and Processing Technologies*, 1–20.

Kumar V., Singh R., & Ahuja I. P. S. (2021c). Comparison of mechanical blended and chemical assisted mechanical blended ABS-Graphene Reinforced Composite for 3D Printing Applications. *Reference Module in Materials Science and Materials Engineering*, doi.org/10.1016/B978-0-12-820352-1.00091-2

Kumar R., Singh R., Kumar V., & Kumar P. (2021d). On Mn doped ZnO nanoparticles reinforced in PVDF matrix for fused filament fabrication: Mechanical, thermal, morphological and 4D properties. *Journal of Manufacturing Processes*, *62*, 817–832.

Kumar V., Singh R., & Ahuja I. P. S. (2021e). On 4D capabilities of chemical assisted mechanical blended ABS-nano graphene composite matrix. *Materials Today: Proceedings*, doi.org/10.1016/j.matpr.2021.05.678.

Kumar V., Singh R., Ahuja I. P. S. *et al.* (2021f). On Nanographene-Reinforced Polyvinylidene Fluoride Composite Matrix for 4D Applications. *Journal of Materials Engineering and Performance*, *30*, 4860–4871. 10.1007/s11665-021-05459-z

Kumar V., Singh R., & Ahuja I. P. S. (2021g). On programming of PVDF-$CaCO_3$ composite for 4D printing applications in heritage structures. *Proceedings of the Institution of Mechanical Engineers, Part L: Journal of Materials: Design and Applications*. InPress.

Sharma R., Singh R., & Batish A. (2020). On effect of chemical-assisted mechanical blending of barium titanate and graphene in PVDF for 3D printing applications. *Journal of Thermoplastic Composite Materials*, 0892705720945377.

Sharma R., Singh R., Batish A. & Ranjan N. (2021). Investigations on Chemical Assisted Mechanically Blended 3D Printed Functional Prototypes of PVDF-BaTiO3-Gr Composite. *Reference Module in Materials Science and Materials Engineering*, doi.org/10.1016/B978-0-12-820352-1.00144-9

Singh M., Kumar S., Singh R., Kumar R., & Kumar V. (2021). On shear resistance of almond skin reinforced PLA composite matrix-based scaffold using cancellous screw. *Advances in Materials and Processing Technologies*, 1–24.

Singh R., & Singh N. (2017). Effect of hybrid reinforcement of SiC and Al2O3 in Nylon-6 matrix on mechanical properties of feed stock filament for FDM. *Advances in Materials and Processing Technologies*, *3*(3), 353–361.

7 Economic and Environmental Justification

Ranvijay Kumar and Rupinder Singh

7.1 INTRODUCTION

The recycling of plastic solid waste has become one of the most essential processes for the sustainability of the land living, aquatic, and wildlife species. In line with the recycling of plastic solid waste, the term "circular economy" is the aspect of sustainable manufacturing or recycling process. The circular economy is the economic system for maintaining little or no loss in the value or properties of the product being processed or manufactured. The success of the concept "circular economy" is dependent upon the use of the raw materials or the system change (Liu et al., 2018; Sanchez et al., 2020; Bora et al., 2020). In other words, the success of the circular economy is declared by the sustainable use of the products, materials, and system. The circular economy of plastic-based materials may be maintained by the use of the waste management hierarchy concept. The waste management hierarchy explains that we should limit the use of plastic (prevention) rather than its use. If it has been used in any form, we should think about the reuse of plastics. Subsequently, If it is not possible to reuse those plastics, then we should think of recycling and ultimately if it is not in the scope of recycling then we should finally think about its disposal to recover the energy or fuel (Gertsakis & Lewis, 2008; Pires & Martinho, 2019; Price and Joseph 2000, Eriksson et al., 2015). In this regard, the previous researchers have done many studies for boosting the circular economy by plastic recycling through processes like extrusion, 3D printing, molding, drawing, investment casting, etc. Extrusion is the process of mechanical recycling of plastics that is done near the melting point under the action of extrusion torque.

The qualities of the 3D-printed parts in terms of circular economy are largely dependent upon the processing of the feedstock filaments. So, the extrusion process may be one of the possible considerations for the circular economy in plastic recycling, and the economy and environmental factors must depend upon it. The previous researchers have used reinforcement of the metallic, non-metallic, and oxide particles in the thermoplastic matrix for boosting the circular economy. Several studies have been reported for mechanical recycling by an extrusion process for sustainable manufacturing that boosted the circular economy of thermoplastic-based materials. For example, Singh et al. (2019) have used polyvinyl chloride

DOI: 10.1201/9781003184164-7

(PVC), Fe3O4, and wood dust as reinforcement in polylactic acid (PLA) to tune the mechanical, magnetic, thermal, and morphological properties of the thermoplastic composites. Maintaining the biodegradability of the plastic with reinforcement of agriculture wastes is also the key approach in the circular economy of plastic. Some previous studies have revealed the sustainable manufacturing of rice-straw, almond skin powder, and wood dust reinforced with thermoplastics materials by the screw extrusion process for cost-effective production purposes (Osman & Atia, 2018; Kumar et al., 2019a; Singh et al., 2019). The reinforcement has been termed the key role maker to sustain against any type of physical loading. Maintaining the flow of plastic materials can also be termed under the concept of the circular economy. Singh et al. (2018) have maintained the flow rate of acrylonitrile butadiene styrene (ABS) and PA6 materials by reinforcement of aluminum (Al) metal particles to manufacture the sustainable welding joints by the friction welding approach. Maintaining the melt flow rate of the plastics has ensured improved welding strength (Singh et al., 2018; Kumar et al., 2019b; Kumar et al., 2020). Similarly, Tao et al. (2017) have manufactured the wood flour–reinforced PLA thermoplastic feedstock filaments and concluded that the microstructures of the composites materials have been modified. Similar studies have been conducted in the recent past for the development of plastic-based composites materials for tuning the mechanical, morphological thermal, chemical, and wear properties. For example, polycaprolactone (PCL) in PLA, iron in ABS, carbon nanotubes (CNT) in ABS, CNT in polyether ether ketone (PEEK), TiO2 in polyamide 6 (PA6), and Al and Al2O3 in nylon thermoplastic have been used to maintain the circular economy of the products (Haq et al., 2017; Mostafa et al., 2009; Ning et al., 2015; Berretta et al., 2017; Singh et al., 2018; Boparai et al., 2016).

The previous literature shows that the circular economy and the sustainability aspects of the products are largely dependent on the processing route, types of reinforcement, and method of recycling in the case of plastics. The environmental and economic factors are also influenced by the action of the processing parameters and the processing used. In this regard, the process of recycling should be done not only based on a sustainability point of view but the environmental and economic points of view also. So, in this chapter, detailed literature and discussions have been presented for keeping the environmental and economic aspects of the circular economy of recycled plastics.

7.2 BACKGROUND FOR CIRCULAR ECONOMY IN PLASTIC RECYCLING

The circular economy of plastic-based products/materials is affected by the method of recycling and their impact on the environment. In other words, the environmental and economic justifications of the circular economy are dependent upon the properties of the finally developed products. To investigate the background of the circular economy concerning plastic recycling and environmental impact, the analysis has been conducted by using a database available on www.webofknowledge.com. On inputting the keywords "plastics recycling, environment, circular economy," there is a total of 274 research papers that have been

TABLE 7.1
Terms, their occurrences, and relevance score for plastic recycling, environment, and circular economy

No.	Term	Occurrences	Relevance Score
1	Polymer	44	0.2582
2	Life	34	0.3867
3	Emission	30	0.2752
4	Pollution	30	0.7306
5	Sustainability	27	0.4491
6	Environmental impact	26	0.4602
7	Energy	25	0.4847
8	Market	25	0.3839
9	Implementation	24	0.5248
10	Metal	24	1.1909
11	Polyethylene	24	1.859
12	Fuel	23	0.6293
13	Context	22	0.6375
14	Life cycle assessment	22	0.5727
15	Incineration	21	0.5958
16	Polypropylene	20	1.9994
17	Circularity	19	0.6237
18	Municipal solid waste	18	0.6759
19	Polyethylene terephthalate	18	1.0832
20	Ocean	17	1.2598
21	Term	17	0.5414
22	Waste plastic	17	0.2986
23	Recycled plastic	16	2.5792
24	Microplastic	15	2.007
25	Plastic pollution	15	1.7282
26	Plastic recycling	15	2.1034
27	Energy recovery	14	0.634
28	Food packaging	14	0.5493
29	Electronic equipment	13	2.6089
30	Human health	13	1.3035
31	Environmental benefit	12	1.2672
32	Sustainable development	12	1.2989

found in the years 2013–2021. Out of these 274 research papers, there are 131 relevant terms that have been found by selecting 12 minimum numbers of occurrences of each term. About 60 of the most relevant terms have been kept for analysis and out of those only 32 useful terms have been retained for making network-based diagrams (see Table 7.1).

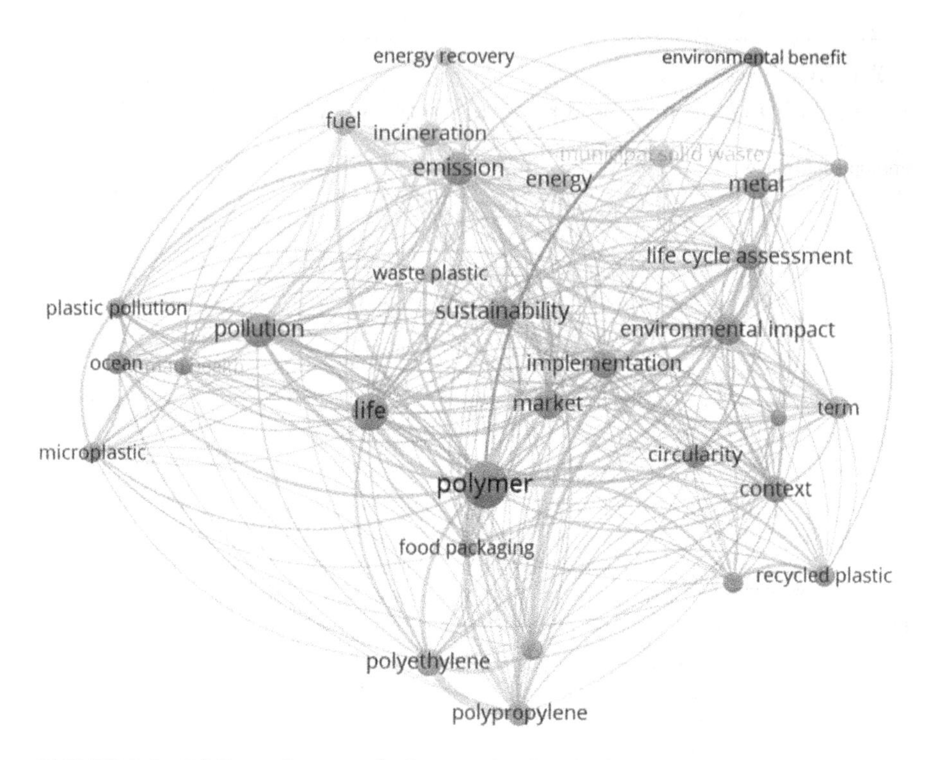

FIGURE 7.1 Bibilometric network diagram for the plastics recycling, environment, and circular economy.

Figure 7.1 shows the network diagram for the most used terms that reflect the relationship between the investigations of the previous studies. There are five different clusters of studies that have been found (in a different color of the node).

As per the relationship developed, it has been observed that most of the previous studies have been related to polymer recycling and its investigation is related to sustainability, implementation of recycling techniques, circularity, pollution concerns, investigations of polyethylene and polypropylene like plastics, its environmental impacts, and energy recovery. Figure 7.2 shows the bibliometric network diagram for plastic recycling, environment, and circular economy in relationship to their environmental impact. The studies revealed tools, techniques, methodology, and processing related to energy recovery, incineration, fuel, environmental benefits, food packaging, and recovery of waste plastics. But hitherto, limited studies have been reported for the life cycle assessment, recovery of ocean waste, circular economy, etc. for the plastic- and micro-plastic-based products (see Figure 7.2). In the upcoming research, the study may be extended for the investigations of the economic and environmental justification of plastic recycling with establishing a circular economy.

There is a huge scope that has been granted to scientists and researchers working on maintaining the circular economy in plastic recycling. This has been confirmed

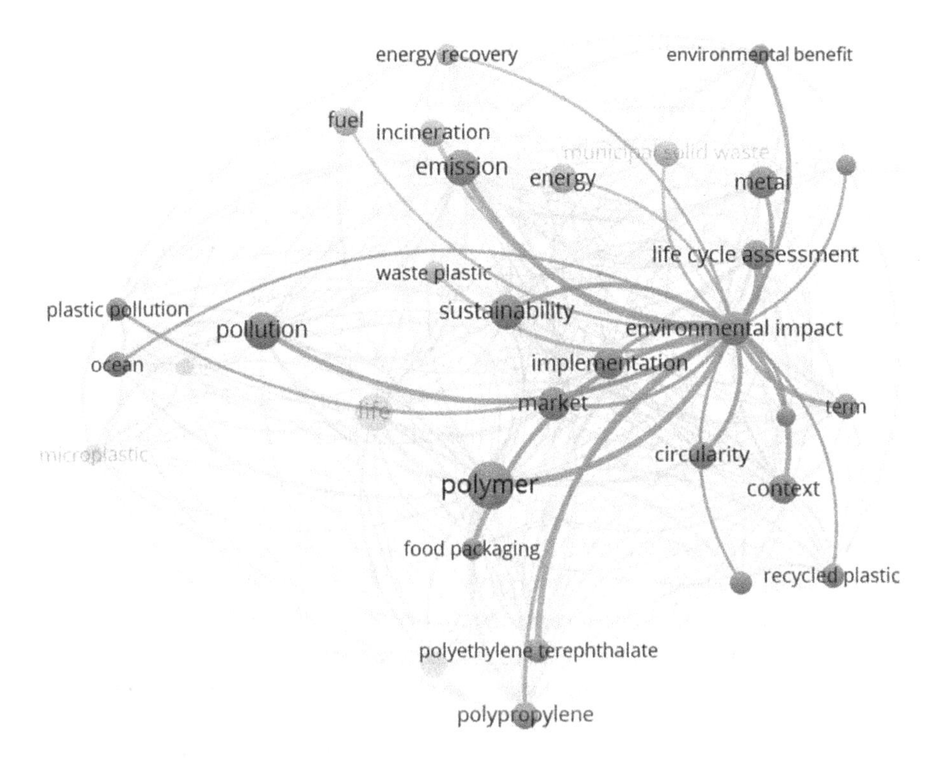

FIGURE 7.2 Bibilometric network diagram for the plastic recycling, environment, and circular economy in relationship to their environmental impact.

by the research publication reported in the previous eight years (Figure 7.3). It has been observed that the research publications for plastic recycling with a caring environment and circular economy has increased year to year and it is expected that there will be a huge scope for researching this area.

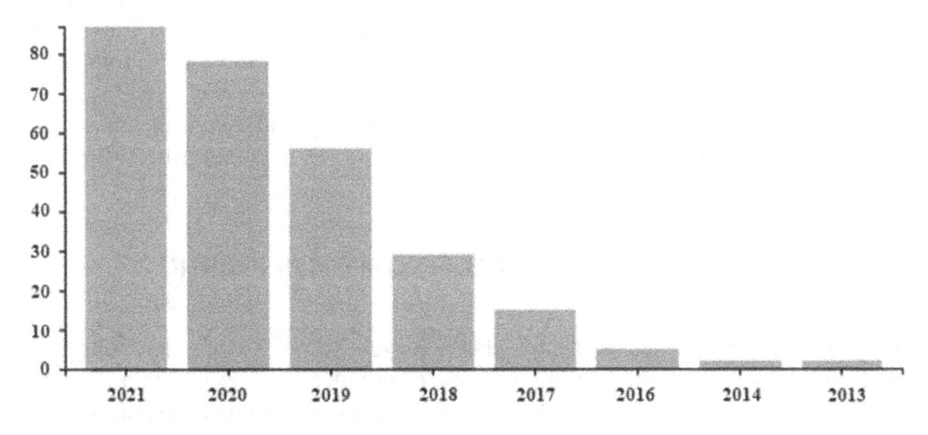

FIGURE 7.3 Year-wise research publications for research related to plastic recycling, environment, and circular economy.

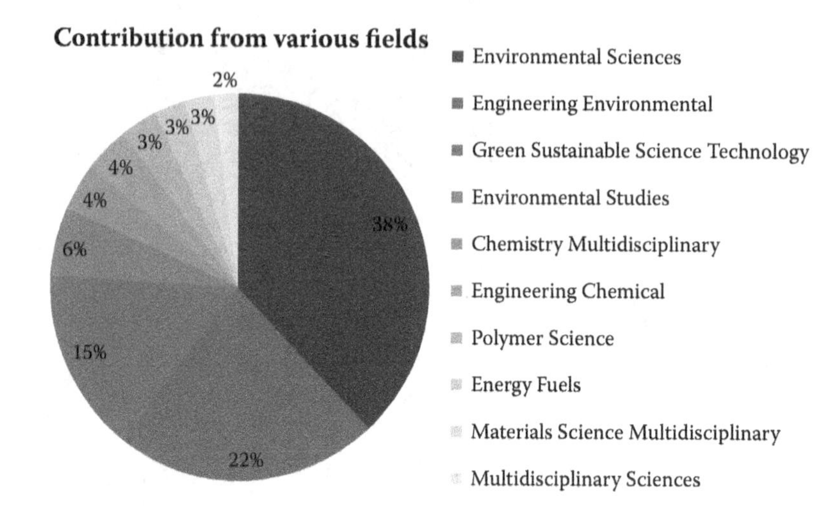

FIGURE 7.4 Research contributions from different areas of studies.

In the recent times, every possible effort has been implemented by the researchers, scientists, government bodies, funding agencies, and NGOs for the research on the circular economy of plastic recycling with the care of environmental and economic justifications.

In this relation, the environmental science field is the most contributing field, which has contributed 38% of total studies followed by environmental engineering (22%), green sustainable science technology (15%), environmental studies (6%), chemistry multidisciplinary (4%), chemical engineering (4%), polymer sciences (3%), energy fuels (3%), multidisciplinary materials science (3%), and multidisciplinary sciences (2%) (see Figure 7.4). Future studies may be performed by collaborating the two different fields for better environmental and economic justification of plastic recycling with maintaining a circular economy. In this regard, there are several funding agencies working positively to reduce plastic waste and maximize the circular economy of plastics. The European Commission is the top funding agency that has contributed almost 21%, followed by the National Natural Science Foundation of China (NSFC) (19%), UK Research Innovation (UKRI) (17%), Engineering Physical Sciences Research Council (EPSRC) (10%), U.S. Department of Energy (DOE) (7%), BMDW (6%), and others (see Figure 7.5).

7.3 THE VISION OF CIRCULAR ECONOMY FOR PLASTIC RECYCLING

The aim of the circular economy for plastic recycling consists of prioritizing regenerative resources, increasing the life span of products, converting waste into a resource, rebuilding the business model, incorporating digital processes, designing for the future, and boosting the recycling strategies (see Figure 7.6).

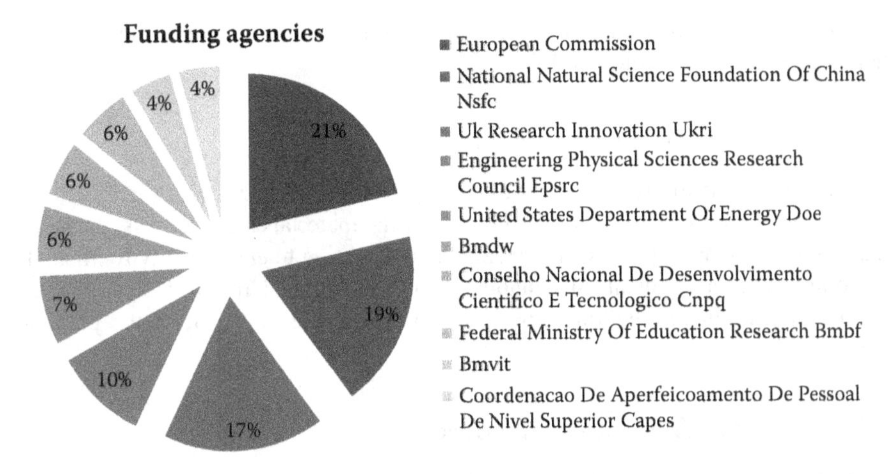

FIGURE 7.5 Leading funding agencies working for the environmental aspect of circular economy in plastic recycling.

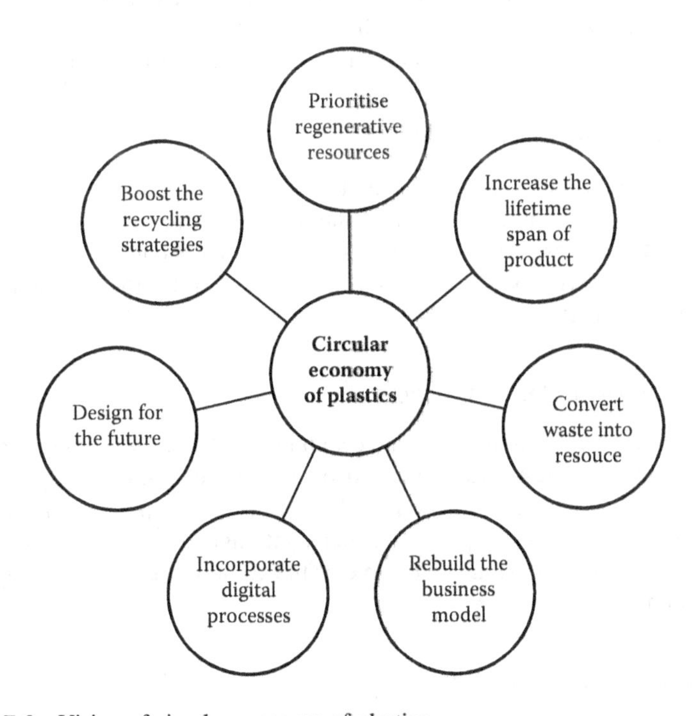

FIGURE 7.6 Vision of circular economy of plastics.

7.3.1 Prioritize Regenerative Strategies

Prioritizing the regeneration of materials, processing, and products is the key aim of the circular economy. In case of plastic recycling, it deals with reusing plastics and plastic-based products by processing and reprocessing. The ultimate regeneration is to make the plastic materials or plastic-based products sustainable as per environmental as well as an economic basis.

7.3.2 Increase the Life Span of the Product

The vision behind the regeneration of the materials and products is to stretch the lifetime For the maximization of the lifetime of the materials and products, the steps that can be taken are reprocessing, reusing, and recycling to enhance environmental, economic, and mechanical sustainability. Introducing and reintroducing plastic-based materials in production cycles many times is the key goal of the circular economy.

7.3.3 Convert Waste Into a Resource

Converting waste materials or products into sustainable and eco-friendly resources is the primary aim of the circular economy. In this way, plastic recycling processes in various ways give a boost to the improvement in the circular economy. The plastic-based waste materials and products can be converted into resources by redesigning, reproduction, remanufacturing, recycling, reusing, and regeneration processes.

7.3.4 Rebuild the Business Model

The most common economic business models, where the economic aspects are kept prioritized over the social and environmental aspects, in the circular economy care for both the business as well as consumers. The circular economy systems provide an effective and better module for the business by caring for the economic as well environmental aspects.

7.3.5 Incorporate Digital Processes

Digital technologies allow the circular economy model to be implemented with better efficiency for plastic recycling. Also, digital technologies help in the re-designing, rebuilding, transformation, and creation of complex business models help in circular supply chains, etc. Artificial intelligence, virtual reality, digital twins, and the Internet of Things are some of the components that provide a better circular economy.

7.3.6 Design for the Future

One of the major advantages of using the circular economy for plastic recycling is it cares for the future aspects of the environment, industry, materials, and products for their sustainable use. The circular economy believes in the eco-design of the products that can sustain up to the maximum life span with minimum losses to the environment, society, and economy. For example, reducing plastic wastes by re-processing is an example of restricting its harmful impact in the future. The circular economy also promotes energy from renewable resources, eco-design, industrial and territorial ecology, and effective use of plastics and plastic-based products for future aspects.

7.3.7 Boost the Recycling Strategies

The effectiveness of plastic recycling is maximized when a sustainable reprocessing and regeneration approach is applied. The circular economy aims to endure the life span of materials and products. For example, in the circular economy framework, the use of plastic-based products is advised to reduce; if it is not possible to reduce, it is advisable to reuse; further, it is not possible to reuse, then it can be recycled. After the end of the recycling phase, the energy recovery processes may be applied to its sustainable and easy dumping. So, based upon the idea, it can be said the circular economy is an effective thought to apply for better implications of plastic recycling.

7.3.8 The Economic Impact of Plastic Recycling

Plastic recycling promotes not only reducing the unwanted pollution loading the earth but at the same time, plastic recycling also leads to creating jobs, creating social benefits in money and wealth as well as improving the economy. In this regard, the price of the polymers is controlled by the various aspects by which the benefits may be gained. The possible aspects of benefits are as follows:

- The export of the recycled materials in the developing and emerging counties where the requirement is very high for polymers, such as packaging stuff.
- There is high demand for high-quality plastics that has gone through recycling. For example, the recovered biocompatible and biodegradable plastics are high in demand.
- The economies are also improved by maximizing the sorting and collection system used for the recycling processes.
- Making more flexible, cleaned, and colorful plastics obtained from the recycling processes help in economic benefits.

As per the study conducted by the Institute of Scrap Recycling Industries (ISRI), approximately 463,000 jobs were provided alone in the USA. The opportunity areas are the recycling of metals, plastics, electronics, rubber, paper, and textile-based products. The report has been suggested that 29.8% (138,000) of direct employment has been provided for recycling, 28.5% (132,000) in supplier jobs, and nearly 41.7% (193,000) in other forms. It has been expected that there should be an increase of 10% in jobs in the next five years (Northeast recycling council (NERC), 2013). Ultimately, plastic recycling helps in contributing the economic benefits in various forms, including the following aspects:

- **Benefits in the government services:** Transportation, public health and safety, and social service programs.
- **Benefits from recycled products:** The newly emerging technologies, such as 3D printing, are highly efficient in generating revenue from recycled products.
- **Benefits in imports/exports:** Sales of recycled goods, selling transformed energy, coal, cotton, and ethanol obtained from recycling.
- **Benefits in expanding local employer opportunities:** Generating wealth from recycling tax revenue.
- **Benefits by energy recovery:** The expenditure of the billions of dollars may be reduced by adopting the plastic recycling process as energy recovery.

The scientists, researchers, and politicians should work on the development of the processes, strategies, and the law for the development of an efficient and eco-friendly recycling approach that can sustain the volatile, dynamic, and globalizing world. In this direction, the regulating bodies are also working on each and every aspect of the recycling process to get the benefits. In 2000, the World Business Council for Sustainable Development (WBCSD) has proposed that recyclability should be improved as the critical aspect for the sustainable development of businesses (Ferreira et al., 2008). So, for the above-mentioned reason, plastic recycling with maintaining the circular economy should be promoted effectively and efficiently for increasing economic benefits.

7.4 ENVIRONMENTAL IMPACT OF PLASTIC RECYCLING

Worldwide, a total amount of 6.3 billion tonnes of plastics were being produced from 1950 to 2018, among which an amount of 9% has been recycled and 12% of the plastics have been incinerated (Alabi et al., 2019). The demand for plastic is increasing as the population is increasing, which is responsible for the generation of pollution by waste plastic accumulation. Recycling methods are used for years to control the accumulation of waste plastic materials in any form. To control and manage plastic waste, various recycling strategies have been developed such as the following:

- **Primary recycling:** The mechanical processes such as molding, extrusion, heat processing, pultrusion, forming, etc.
- **Secondary recycling:** The mechanical processes such as molding, extrusion, heat processing, pultrusion, forming, etc. with reinforcement and addition of the additive materials for improving the sustainability.
- **Tertiary recycling:** The tertiary recycling processes are employed for those plastics or plastic-based products that cannot be done by the secondary recycling processes. Tertiary recycling is a chemical-based process in which chemical reactions are carried out to re-form the plastics. The chemical-based solvent extraction processes are an example of the tertiary recycling process.

- **Quaternary recycling:** When there is no scope for the waste plastics or plastic-based products to be recycled by any means, then those plastics materials are then incinerated or burned by the controlled combustion process to recover the energy.

As per the pollution concern, primary recycling is a less pollution-generating process, since it works on the virgin or unprocessed types of plastics. Generally, the primary recycling processes are employed to enhance the properties in the form of mechanical, thermal, and molecular properties. The chances of emitting pollution with secondary recycling processes are quite bigger than the primary recycling. The additions of additives make the processing at a higher temperature, which promotes more degradation of the plastics. The tertiary recycling process offers no heat degradation of the plastics but at the same time uses chemicals that may lead to the pollution of the environment. The chemical used in the recycling process is hazardous and increases the possibility of environmental risks. The quaternary recycling process is the most dangerous threat to the environment as the incineration of plastic leads to the emission of dangerous gases into the atmosphere. The exposure of emitted gases from the incineration process is the leading pollution-emitting cause.

As per the environmental benefit concerns of the plastic recycling process, the following are aspects of plastic recycling to the environment (Stanford, 2021):

- **Conservation of energy:** It has been established that the manufacturing/production of the products using the recovered materials usually takes less energy and results in the burning of fewer fossil fuels.
- **Reduction in air and water pollution:** The recycling of plastic is one of the most important aspects for reducing the chances of air and water pollution.
- **Reduction in the emission of greenhouse gases:** The chances of emission of greenhouse gases such as CO_2, CH_4, NO, and chlorofluorocarbons can be reduced by plastic recycling.
- **Conservation of the natural resources:** The natural resources in form of the trees, metal ores, minerals, and other materials used for the manufacturing of everyday use essentials, but the use of recycled plastic may directly restrict or reduce of use of these natural resources.

7.5 THE FUTURE ASPECT OF BOOSTING THE CIRCULAR ECONOMY FOR PLASTIC RECYCLING

Plastic recycling is one of the most helpful ways to maintain the circular economy of the process materials and products. Approaches have been implemented in the past to reduce, reuse, and recycle plastics and plastic-based products. But hitherto, the future is open to boost the circular economy by the plastic recycling processes. Following are the future aspects of plastic recycling for boosting the circular economy:

- **Non-mechanical recycling:** The non-mechanical recycling under the umbrella term "chemical recycling" is expected to contribute approximately 7% for the purification purpose by the end of 2030 (Letsrecycle, 2021). These provisions allow restricting or reducing the landfill process for protecting the environment.
- **Additive manufacturing process:** Additive manufacturing has grown so rapidly in the last decade that it covered almost all areas of applications. Additive manufacturing itself has been termed the technology of the future. The various additive manufacturing processes, such as fused deposition modeling (FDM) and stereolithography (SLA), mostly use plastic-based materials for manufacturing. These processes are also available to recycled, reprocessed, and waste plastics so that there are huge changes to get the economic and environmental benefits from the additive manufacturing processes.
- **Energy recovery:** The accumulation of non-biodegradable plastic wastes that are not possible for either mechanical or chemical recycling are available for the energy-recovery process. In the future, more energy will be recovered from non-biodegradable plastics.
- **Business model:** Plastic recycling is said to be effective only if it contributes to economic growth. The economic growth of the country is dependent upon how the business model carryies out the recycling process. In the future, the recycling components like segregation, collection, shredding, and mechanical recycling will be promoted by the government and industries to get the financial benefits from recycling. The production of essentials and goods for export purposes also comes under this future component.

7.6 SUMMARY

The following is a summary of this chapter:

- The aim of the circular economy for plastic recycling consists of prioritizing regenerative resources, increasing the life span of products, converting waste into a resource, rebuilding the business model, incorporating digital processes, designing for the future, and boosting the recycling strategies.
- The effectiveness of plastic recycling is maximized when the sustainable reprocessing and regeneration approaches are applied. The circular economy aims to ensure the life span of materials and products.
- The approaches implemented in the past accommodate reducing, reusing, and recycling plastics and plastic-based products. But hitherto, the future is all open to boost the circular economy by plastic recycling processes. In the future, promoting non-mechanical recycling, recycling by additive manufacturing, energy recovery, and the business model should be developed for improving the circular economy of plastic recycling.

ACKNOWLEDGEMENTS

The authors are very thankful to the University Center for Research and Development, Chandigarh University and Center for Manufacturing Research, Guru Nanak Dev Engineering College, Ludhiana for their assistance.

REFERENCES

Alabi O. A., Ologbonjaye K. I., Awosolu O., & Alalade O. E. (2019 Apr 5). Public and environmental health effects of plastic wastes disposal: A review. *Journal of Toxicology and Risk Assessment*, *5*(021), 1–3.

Berretta S., Davies R., Shyng Y. T., Wang Y., & Ghita O. (2017 Oct 1). Fused Deposition Modelling of high temperature polymers: Exploring CNT PEEK composites. *Polymer Testing*, *63*, 251–262.

Boparai K. S., Singh R., & Singh H. (2016 Mar 21). Experimental investigations for development of Nylon6-Al-Al2O3 alternative FDM filament. *Rapid Prototyping Journal*, 10.1108/RPJ-06-2014-0076

Bora R. R., Wang R., & You F. (2020 Oct 16). Waste polypropylene plastic recycling toward climate change mitigation and circular economy: energy, environmental, and technoeconomic perspectives. *ACS Sustainable Chemistry & Engineering*, *8*(43), 16350–16363.

Eriksson M., Strid I., & Hansson P. A. (2015 Apr 15). Carbon footprint of food waste management options in the waste hierarchy–a Swedish case study. *Journal of Cleaner Production*, *93*, 115–125.

Ferreira, B., Monedero, J., Martí, J. L., Aliaga, C., Hortal, M. & López, A. D., (2008). The economic aspects of recycling. Edited by Enri Damanhuri, *Post-Consumer Waste Recycling and Optimal Production*, Ch. 6, pp. 99–127.

Gertsakis J., & Lewis H. (2008). Sustainability and the waste management hierarchy. *Retrieved on January. 2003* Mar; 30.

Haq R. H., Rahman M. N., Ariffin A. M., Hassan M. F., Yunos M. Z., & Adzila S. (2017 Aug 1). Characterization and mechanical analysis of PCL/PLA composites for FDM feedstock filament. In *IOP Conference Series: Materials Science and Engineering* (*Vol. 226*, No. 1, p. 012038). IOP Publishing.

Kumar R., Singh R., Ahuja I. P., & Fortunato A. (2020 Dec 1). Thermo-mechanical investigations for the joining of thermoplastic composite structures via friction stir spot welding. *Composite Structures*, *253*, 112772.

Kumar R., Singh R., & Ahuja I. P. (2019b Jan 1). Friction stir welding of ABS-15Al sheets by introducing compatible semi-consumable shoulder-less pin of PA6-50Al. *Measurement*, *131*, 461–472.

Kumar S., Singh R., Singh T. P., & Batish A. (2019 Sep). Investigations of polylactic acid reinforced composite feedstock filaments for multimaterial three-dimensional printing applications. *Proceedings of the Institution of Mechanical Engineers, Part C: Journal of Mechanical Engineering Science*, *233*(17), 5953–5965.

Kumar S., Singh R., Singh T. P., & Batish A. (2019a Jun 11). On investigation of rheological, mechanical and morphological characteristics of waste polymer-based feedstock filament for 3D printing applications. *Journal of Thermoplastic Composite Materials*, 0892705719856063.

Letsrecycle, https://www.letsrecycle.com/news/what-is-the-future-of-plastic-recycling/, retrieved on 29th September 2021.

Liu Z., Adams M., Cote R. P., Chen Q., Wu R., Wen Z., Liu W., & Dong L. (2018 Aug 1). How does circular economy respond to greenhouse gas emissions reduction: An analysis of Chinese plastic recycling industries. *Renewable and Sustainable Energy Reviews*, *91*, 1162–1169.

Mostafa N., Syed H. M., Igor S., & Andrew G. (2009 Jun 1). A study of melt flow analysis of an ABS-Iron composite in fused deposition modelling process. *Tsinghua Science & Technology*, *14*, 29–37.

Ning F., Cong W., Qiu J., Wei J., & Wang S. (2015 Oct 1). Additive manufacturing of carbon fiber reinforced thermoplastic composites using fused deposition modeling. *Composites Part B: Engineering*, *80*, 369–378.

Northeast recycling council (NERC). (2013). Economic Impact of Recycling, https://nerc.org/news-and-updates/blog/nerc-blog/2013/10/08/economic-impact-of-recycling, retrieved on 28th September 2021.

Osman M. A., & Atia M. R. (2018 Aug 13). Investigation of ABS-rice straw composite feedstock filament for FDM. *Rapid Prototyping Journal*, *24*, 1067–1075

Pires A., & Martinho G. (2019 Jul 15). Waste hierarchy index for circular economy in waste management. *Waste Management*, *95*, 298–305.

Price J. L., & Joseph J. B. (2000 May). Demand management–a basis for waste policy: a critical review of the applicability of the waste hierarchy in terms of achieving sustainable waste management. *Sustainable Development*, *8*(2), 96–105.

Sanchez F. A., Boudaoud H., Camargo M., & Pearce J. M. (2020 Aug 10). Plastic recycling in additive manufacturing: A systematic literature review and opportunities for the circular economy. *Journal of Cleaner Production*, *264*, 121602.

Singh R., Kumar R., & Ahuja I. P. (2018 Nov 12). Mechanical, thermal and melt flow of aluminum-reinforced PA6/ABS blend feedstock filament for fused deposition modeling. *Rapid Prototyping Journal*, *24*, 1455–1468

Singh R., Kumar R., Mascolo I., & Modano M. (2018 Jun 15). On the applicability of composite PA6-TiO2 filaments for the rapid prototyping of innovative materials and structures. *Composites Part B: Engineering*, *143*, 132–140.

Singh R., Kumar R., Pawanpreet, Singh M., & Singh J. (2019 Nov 5). On mechanical, thermal and morphological investigations of almond skin powder-reinforced polylactic acid feedstock filament. *Journal of Thermoplastic Composite Materials*, 0892705719886010.

Stanford, PSSI /Stanford recycling land, building and real estate, https://lbre.stanford.edu/pssistanford-recycling/frequently-asked-questions/frequently-asked-questions-benefits-recycling, retrieved on 29th September 2021.

Tao Y., Wang H., Li Z., Li P., & Shi S. Q. (2017 Apr). Development and application of wood flour-filled polylactic acid composite filament for 3D printing. *Materials*, *10*(4), 339.

8 Twin-Screw Extrusion for Processing Thermoplastics in Biomedical Scaffolding Applications

Nishant Ranjan, Rupinder Singh, Ranvijay Kumar, and Ravinder Sharma

8.1 INTRODUCTION

Extruders (TSE and SSE) are the most demandable machines for the processing of food, metals, and plastic industries and the use of extrusion procedures is especially common in product manufacturing that uses polymers as raw material (Kumar et al., 2020). Extruded polymer goods commonly include films, hoses, sheets, insulated wires, cables, pipes, and tiles (Ramteke et al., 2017; G. Singh et al., 2019). According to workings and principle, mainly extruders are normally categorised as SSE or TSE. 3D-printing technology is one of the fast-growing technologies in manufacturing sectors due to being easy to fabricate any size and ease of available input materials. As per different applications, with the former being frequently used in general polymer processing and the latter for compounding various fibres, fillers, and polymer mixes before final moulding. Intermeshing and non-intermeshing TSEs are two types of TSE based on the interactions of the two screws (Batoo et al., 2017; Mousa & Heinrich, 2010; R Singh et al., 2019). Because of their positive displacement features, completely intermeshing counter-rotating TSE have been discovered to offer the best pumping capacities in the TSE family (Gatenholm et al., 1993; Huneault et al., 1992; Pita et al., 2002). In today's industry, TSE is frequently used (Braun, 2001). Co-rotating and counter-rotating TSE are two varieties of such extruders based on the relative rotational direction of their screws (Franzen et al., 2009; Ha et al., 1998; Kim et al., 1999). The maximum velocity in a co-rotating TSE is located at the screw tips, whereas the maximum velocity in a counter-rotating TSE is found in the intermeshing zone (Canel et al., 2012; Singh & Ranjan, 2018; Zhu et al., 2015). It may be argued that because the material is moved between the lobes, the co-rotating mechanism promotes greater mixing (Ranjan et al., 2019a; Ranjan et al., 2021). The counter-rotating mechanism, on the other hand,

DOI: 10.1201/9781003184164-8

provides higher pressure, making profile extrusion more efficient (Kim et al., 1999; Ranjan et al., 2021).

TSEs are used to process virtually every type of thermoplastic polymers that are used today (Ognedal et al., 2012; Ranjan, 2021; Ranjan et al., 2021). The goal of this chapter is to define how TSEs can be used to process thermoplastic polymer formulations that contain additives and fillers into pellets and shapes for use in different sectors, especially in the biomedical area (Crespo et al., 2008; Lertwimolnun & Vergnes, 2006; Ranjan et al., 2021). The TSE is used for processing/reinforcement of thermoplastic polymers using different fillers as per different applications in their respective field. TSEs are normally used for mixing two or more than two polymers with different specifications at a homogeneous rate and that's why its design is also an important factor (Ognedal et al., 2012; Ranjan, 2021). TSEs have high specific energy inputs, high mixing efficiency, and the capacity to process solids and materials with a high viscosity (Crespo et al., 2008; Lee & Han, 2000). As a result, they're appealing to a variety of businesses, including the chemical and pharmaceutical industries, for compounding, reacting, and blending. Additionally, these models can be used to improve screw setup and operating conditions (Meijer & Elemans, 1988; Ranjan et al., 2021; XU et al., 2004).

Biomaterial research in TE and regenerative medicine is focused on the design and manufacture of scaffolds (Aid et al., 2017; Haghshenas, 2016). Over the last two decades, a lot of work has gone into improving potentially useful scaffold materials for TE (Silva & Rezende, 2013; S. Singh et al., 2019). These scaffolds provide mechanical support and facilitate cell proliferation and differentiation throughout neo-tissue creation in vitro, as well as during the initial phase and after implantation (Hage et al., 1999; Singh et al., 2017). Polymers with synthetic origins, such as polylactic acid (PLA), polyvinyl alcohol (PVA), poly (glycolic acid) (PLGA), polyglycolic acid (PGA), poly(z-caprolactone) (PCL), poly(ethers) containing poly (ethylene oxide) (PEO), and polyurethane (PU), polyvinyl alcohol (PVA), poly (ethylene glycol) (biomaterials found in nature, such as polypeptides and polysaccharides, are also investigated; (Ben Difallah et al., 2012; Sudeepan et al., 2014). Composites or blends of these synthetic or natural polymers, alone or in combination, can provide a wide range of physicochemical and biological properties (Ranjan et al., 2017, 2019c; Ranjan et al., 2020). A polymer composite with one, two, or three dimensions in various polymer matrices is referred to as a multiphase solid material. In this overview, we look at polymers (both natural and synthetic) as a reinforcement and matrix in a composite (Nazeer et al., 2020; J. Zhu et al., 2020). Mechanical properties, chemical composition, and degradation mechanisms are all used to define scaffold materials. Biomaterial selection is critical in the design and manufacture of medical implants and TE products (Han et al., 2020; Ranjan et al., 2020d; Sathiyavimal et al., 2020). Although the traditional selection criteria for a healthy, long-lasting implant is the use of passive inert material, any artificial material placed in a patient's body causes a cellular response (Ranjan et al., 2020b; Shunmugam & Kanthababu, 2018; J. Singh et al., 2019). As a result, it is now recognised that, rather than simply acting as an inert body, a biomaterial must be biologically appropriate and interact with the tissue when implanted (Deshmukh et al., 2019; Ranjan et al., 2019b). We report on the potential natural and synthetic polymers that have been investigated for many years, as well as their desirable qualities and limitations, in this study piece (Ranjan et al., 2019b; Varma et al., 2020). Furthermore, by overcoming the constraints of each material, the combination of two or more biomaterials with enhanced capabilities in the form of copolymers, new materials have been prepared using blending of polymers with other polymers, and polymers with other fillers that are used in the biomedical area (Asadi et al., 2020). This article discusses a number of important aspects for scaffold design, commercial viability, and fabrication procedures (Nadagouda et al., 2020).

Biomedical polymers continuously play/participate an important role in the field/area of biomedical due to better properties such as biocompatibility, controllable biodegradability, bioactivity, and some other properties such as better mechanical and thermal properties (Ranjan et al., 2019c; Ranjan et al., 2020c; Rayate & Jain, 2018). Due to these superior properties, these polymers are most widely used in biomedical applications, such as drug delivery, fabrication of biomedical instruments/tools, fabrication of biocompatible scaffoldings/implants, and organ transplantations (Sudeepan et al., 2014). Some of the biocompatible and bioactive polymers have better healing properties of wounds and other cuts that have been reinforced with other supportive fillers, such as apatite, chitosan (CS), and other natural biopolymers (Falahati et al., 2020; Ranjan et al., 2020d, 2020a). In the biomedical area, all the instruments that are used during any operations or any work that comes in direct or indirect contact with a human/animal body of any tissue must be biocompatible in nature. These days, biocompatible and biodegradable thermoplastic polymer-based scaffoldings/implants have been fabricated using different manufacturing techniques (Amanat et al., 2010; Falahati et al., 2020; Ranjan et al., 2020c; Rayate & Jain, 2018).

Based on the previous literature study/review, it has been observed that some of the research work has been performed on different thermoplastic polymers on a SSE and TSE but very little has been reported on the processing of different thermoplastic polymers with different fillers by the TSE. According to the literature review, it has been observed that some of the research work has been performed by processing different thermoplastic polymers with SSE but much less has been reported on the reinforcement of different biocompatible and biodegradable thermoplastic polymers with bioactive fillers using the TSE. In this research study, an attempt has been made for the fabrication of biocompatible and biodegradable feedstock filaments with a diameter of 1.75 mm. After fabrication of biocompatible feedstock filaments, 3D scaffoldings/implants have been printed using these feedstock filaments and 3D-printing technology.

8.2 RESEARCH BACKGROUND

8.2.1 Bibliographic Analysis

Bibliographic analysis has been performed for this research study. For this research study, at first three of the most suitable keywords (thermoplastic polymers, TSE, and biomedical applications) have been selected to draw a bibliometric map. These three keywords are put into the www.scopus.com database collections. As per different keywords and their relationship output results in the form of published papers and articles. As per these three different keywords (thermoplastic polymers, TSE, and biomedical applications) searched in the Scopus database collection, an output has been downloaded in ".ris" format. According to the www.scopus.com database, a total of 248 relevant research work is published or accepted. For an analysis of bibliometric analysis, "VOS viewer version 1.6.16" software has been used for developing a detailed bibliometric relationship and research work in different fields/areas. According to downloaded output results of the Scopus database, a total of 3,623 terms have been shown. For this study, for minimizing and selection of the best suitable terms, select the minimum number of occurrences is 5, and a total of 128 terms have been suggested or fulfil the requirements. According to VOS viewer, 60% of the total suggested terms have been filtered and shown as output; for this research study work, 77 (60% of 128) terms were suggested and finally, as per requirements of this research study, 57 terms have been manually selected and 20 terms that are not suitable or less suitable are excluded. Figure 8.1 shows that all 57 terms according to their different fields/areas and four different colours (green, red, yellow, and blue) show four different clusters according to their applications and uses. The red colour mainly shows cluster 1, whereas the green, blue, and yellow colours show clusters 2, 3, and 4, respectively, and have 17, 16, 13, and 11 terms according to their different areas, applications, and requirements. In Table 8.1, the relevance score and their number of occurrences are tabulated.

TABLE 8.1
Biocompatible thermoplastic-related terms, their occurrences, and their relevance score

Serial No.	Terms	Occurrences	Relevance Score
1	Biodegradability	5	0.5119
2	Biodegradable Material	5	1.8595
3	Biodegradable Polymer	14	0.5958
4	Biomedical Field	6	0.5524
5	Biopolymer	8	0.4465
6	Blend	7	0.542
7	Bone	7	0.5702
8	Bone Tissue Engineering	5	0.5331
9	Cell Adhesion	7	0.2112
10	Cellulose	6	0.517
11	Chemical	6	0.9534
12	Collagen	7	0.8152
13	Copolymer	7	0.503
14	Differentiation	11	0.2848
15	Drug	10	0.697
16	Drug Delivery System	8	0.5373
17	Electron Microscopy	10	1.3739
18	Electro Spinning	8	0.585
19	Ethylene Glycol	5	0.8527

TABLE 8.1 (Continued)
Biocompatible thermoplastic-related terms, their occurrences, and their relevance score

Serial No.	Terms	Occurrences	Relevance Score
20	Extracellular Matrix	11	0.7842
21	Fabrication	12	0.7301
22	Fiber	15	0.43
23	Flexibility	6	0.6575
24	Growth Factor	9	0.6761
25	Implant	13	0.7845
26	Implantation	7	1.0744
27	Lactic Acid	10	1.1017
28	Metal	5	1.1781
29	Microenvironment	8	0.7259
30	Molecular Weight	10	1.0635
31	Morphology	11	2.1241
32	Nano-composite	6	0.7181
33	Nanofiber	8	0.9157
34	Nanoparticle	9	0.4182
35	Natural Polymer	6	0.8587
36	Nutrient	7	0.2317
37	Organ	8	0.6465
38	Pla	8	2.7009
39	Plga	5	2.4221
40	Plla	5	3.4458
41	Polylactic Acid	5	3.3651
42	Pore	6	2.5537
43	Porosity	11	0.9198
44	Presence	6	0.8306
45	Production	9	1.4819
46	Regeneration	25	0.1596
47	Regenerative Medicine	9	0.4777
48	Spectroscopy	8	1.8441
49	Stent	6	3.3505
50	Strategy	15	0.4434
51	Synthetic Polymer	6	0.5131
52	Temperature	12	1.4034
53	Tissue Engineering Scaffold	9	0.5243
54	Tissue Engineering Scaffolding	6	0.658
55	Tissue Repair	6	0.6079
56	Tissue Scaffolding	14	0.5846
57	Vitro	8	0.6524

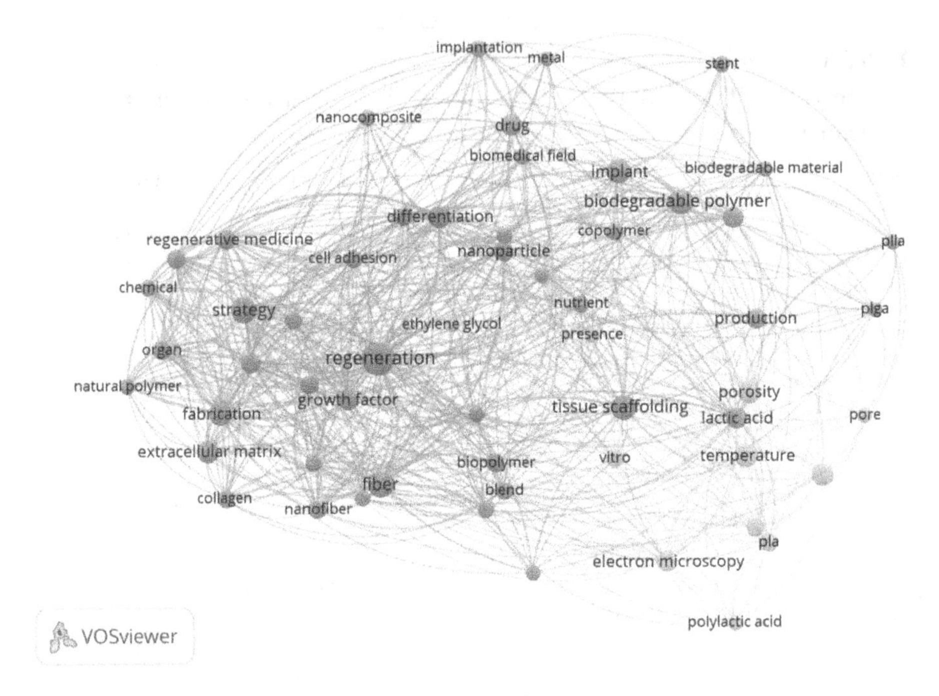

FIGURE 8.1 Bibliographic image for keywords; thermoplastic polymers, TSE, and biomedical applications by using the www.scopus.com database.

According to Table 8.1, it has been observed that regeneration terms have a maximum occurrence (25) and polylactic acid has a maximum relevance score of 3.3651. Based on Table 8.1, Figure 8.1 has been created using VOS Viewer.

Based on Table 8.1 and Figure 8.1, it is observed that lots of research work has been done in the last two decades related to thermoplastic polymers, TSE, and their biomedical applications. For finding research gaps related to the blending of thermoplastic polymers by TSE, one more bibliographic analysis has been performed that shows previous research work done related to them and also shows research gaps related to the blending of thermoplastic polymers with respect to other terms.

According to Figure 8.1 and Table 8.1, Figure 8.2 has been fabricated to determine the relationship among all the terms related to blending of thermoplastic polymers by TSE. Figure 8.2 was created to show the relationship between all the terms concerning the blending of thermoplastic polymers. According to Figure 8.2, it has been concluded that some of the areas, such as biopolymer, lactic acid, regeneration, collagen, bone, PLA, nutrient, biodegradability, implant, drug delivery system, biodegradable materials, etc., have been previously explored related to the blending of thermoplastic polymers. Based on Figure 8.2, the research gap related to this work has been determined, such as extracellular matrix, pore size, bone TE, polylactic acid, PLLA, PLGA, nanocomposite, stunt, copolymer, reinforcement, etc.

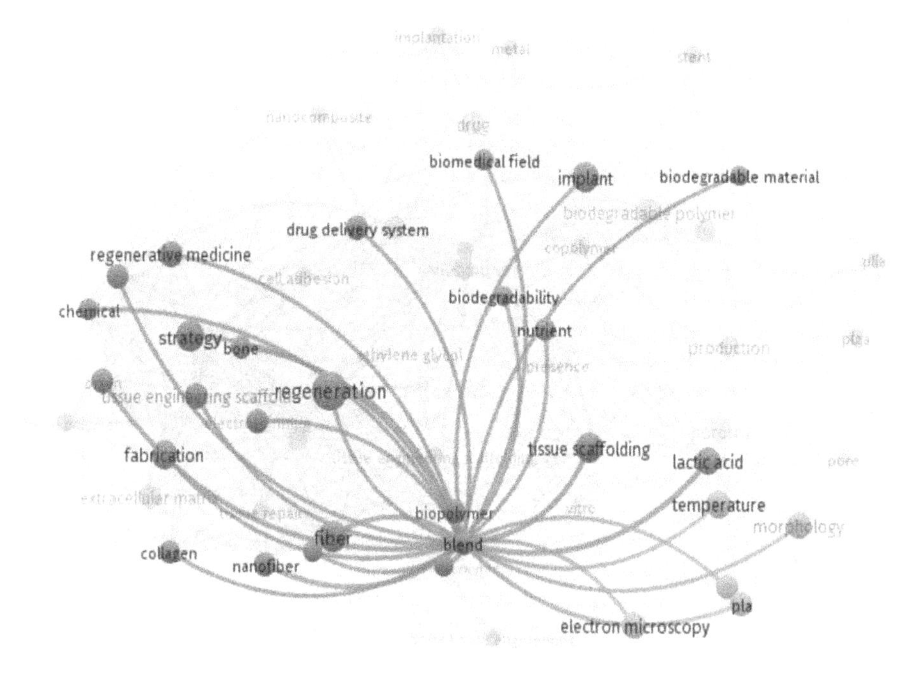

FIGURE 8.2 Bibliographic research gap image for keywords; blending of thermoplastic polymers related to other terms.

8.2.2 Previous Research Work Analysis

For this research study, a different analysis has been performed such as previous research work published linked with the processing of thermoplastic polymers using SSE/TSE according to different year-wise from 1992 till 2021, shown in Figure 8.3. Figure 8.3 shows previous published research work linked with the processing of thermoplastic polymers using TSE/SSE. This research work graph has been created using the www.scopus.com database and the observation has been made that year by year according to published research work related to the processing of thermoplastic polymers is increasing. Maximum research work related to this work is published in 2014 and that is shown in Figure 8.3.

The Scopus database research work related to this study was also analysed according to the subject area and types of documents, shown in Figure 8.4. A pie chart related to this research study according to different subject areas is shown in Figure 8.4(a) and it is observed that the maximum published research work in the area of material science was 27.2% and, after that engineering was 25.1%. In Figure 8.4(b), a pie chart is shown according to different document types and it is observed that research articles are the maximum published based on the Scopus database.

Lastly, this research analysis has been performed according to the research origin country and that is shown in Figure 8.5. According to Figure 8.5, it is observed that

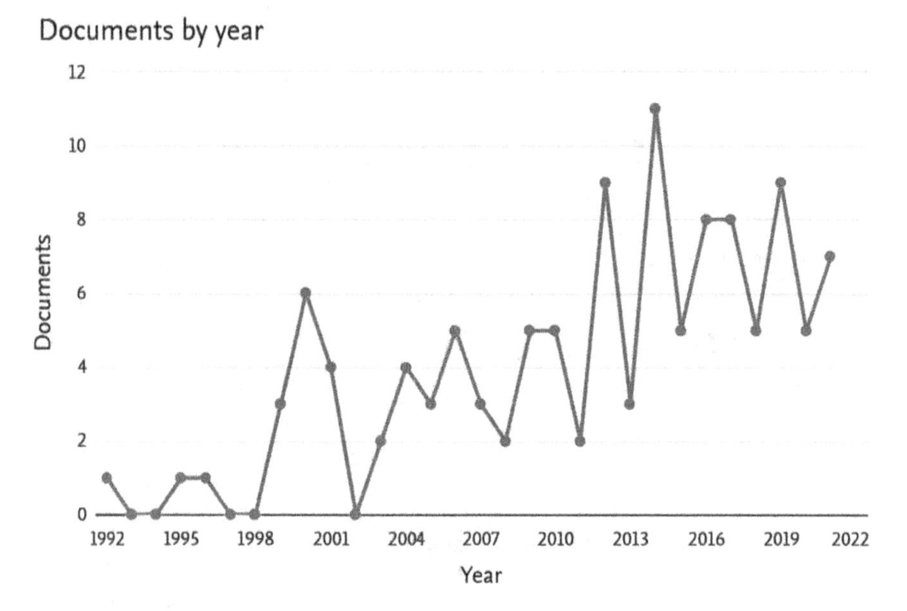

FIGURE 8.3 Last 20 years of research work analysis related to processing thermoplastic polymers using TSE.

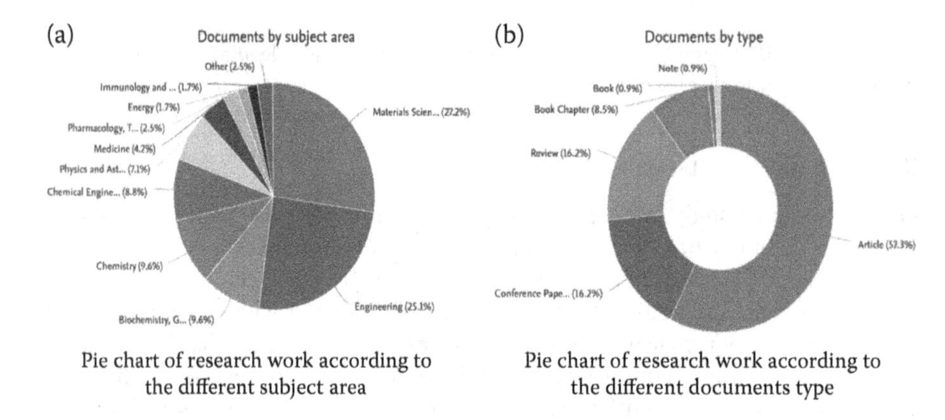

FIGURE 8.4 Pie chart of last 20 years of research work related to processing thermoplastic polymers according to different research areas and document type.

the maximum number of research work documents (58) related to the processing of thermoplastic polymers is published in the United States.

Based on the different studies and research work analysis related to this research study, it has been observed that lots of research work has been done in the last 10 years. The maximum number of research articles published related to the processing of thermoplastic polymers using the TSE are published in the materials science and engineering area by the United States.

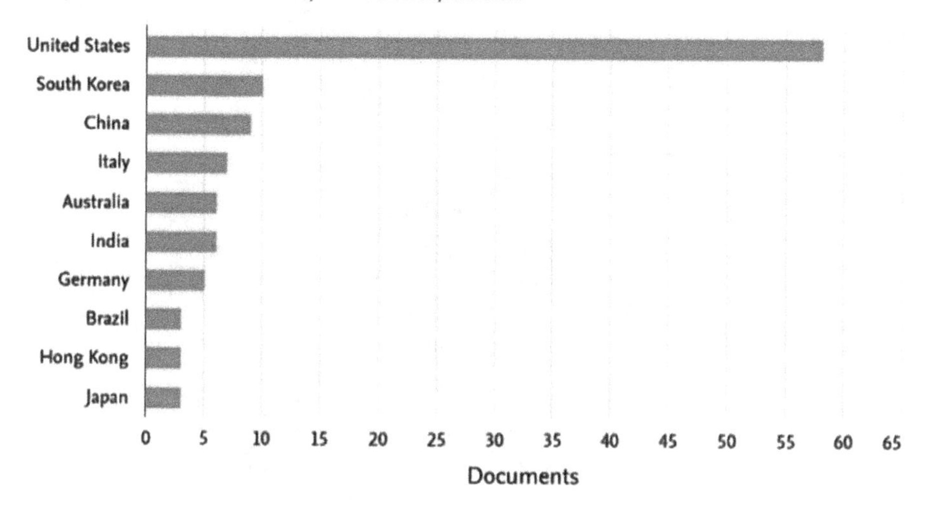

FIGURE 8.5 Published research documents country by country related to this research study.

8.3 BIOMEDICAL APPLICATIONS OF DIFFERENT THERMOPLASTICS

Thermoplastic polymers are used in every field, as per different requirements due to better properties, such as low-cost material, better mechanical strength, lightweight, higher melting temperature, and ease of availability in different sectors (biomedical, aerospace, automobile, construction, mechanical, and chemical). Today, the biomedical field is one of the most important and consumable areas of thermoplastic polymers such as PLA, PLLA, PLGA, PEEK, PP, PE, and PVC. Biocompatible and biodegradable scaffolds and implants are manufactured using polymers that are biodegradable, such as PLA, PLLA, PLGA, and PEEK. Different biomedical applications of biocompatible thermoplastic polymers are shown in Figure 8.6.

Based on previous studies and previous research work, some data has been collected related to different biomedical thermoplastic polymers, their biomedical applications, and properties with their trade name and tabulated in Table 8.2.

8.4 CASE STUDY

8.4.1 Material Processing of PLA-HAp-CS Using TSE for Fabrication of Biocompatible Scaffolds

In this case study, a PLA thermoplastic polymer has been processed with hydroxyapatite (HAp) and chitosan (CS) materials. Ranjan et al. (2019c) have worked on the material processing of biocompatible and biodegradable thermoplastic polymers

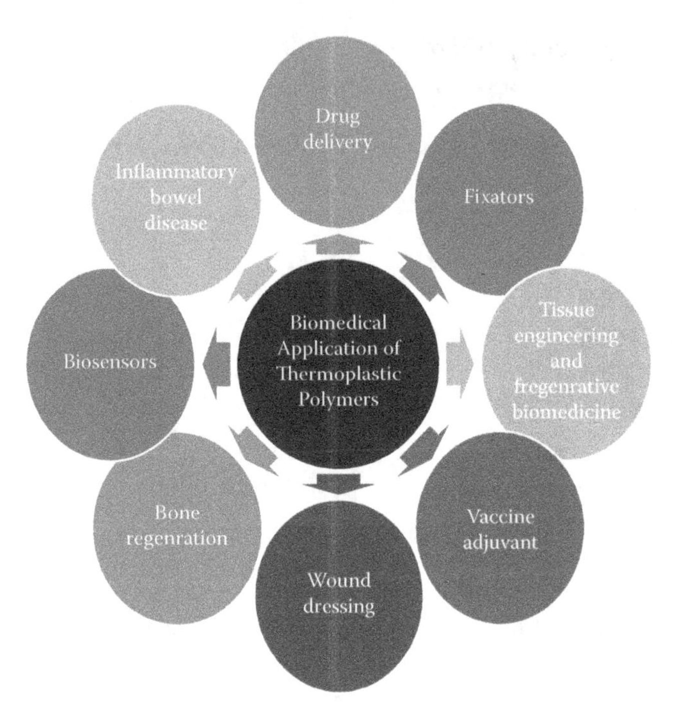

FIGURE 8.6 Different applications of thermoplastic polymers in the biomedical area.

TABLE 8.2
Different biocompatible thermoplastic polymers and their biomedical applications, their property, and trade name

Serial No.	Polymer	Biomedical Application	Property	Trade Name
1.	Collagen	Bioengineered skin equivalents	Scaffolds for musculoskeletal and nervous TE	TransCyte
2.	PLLA	Ligament replacementOrthopaedic devicesFacial fat loss	Injectable form, nondegradable fibers, better tensile strength	DEXONDacron
3.	PLGA	Skin graftMulti-filament structure	Meshes formation, controllable degradation rate	CRYL, Vicryl
4.	PEG	Drug delivery for small and medium-size molecules (active type)	Bioresorbable multiblock	SynBiosys
5.	PLA	Scaffold fabricationImplant fabricationTissue engineering	Better mechanical strength, easy to reinforce	ACRYSTEX MS RESIN
6.	PVC	Single-use applicationsDrug deliveryBiomedical devices	Impermeable to germs, higher elongation and flexibility	Plastisols
7.	PCA	Wound dressingSkin applicationsDrug delivery applications	Better dispersion property, excellent lubricity, excellent thermal and mechanical properties	Carbomer

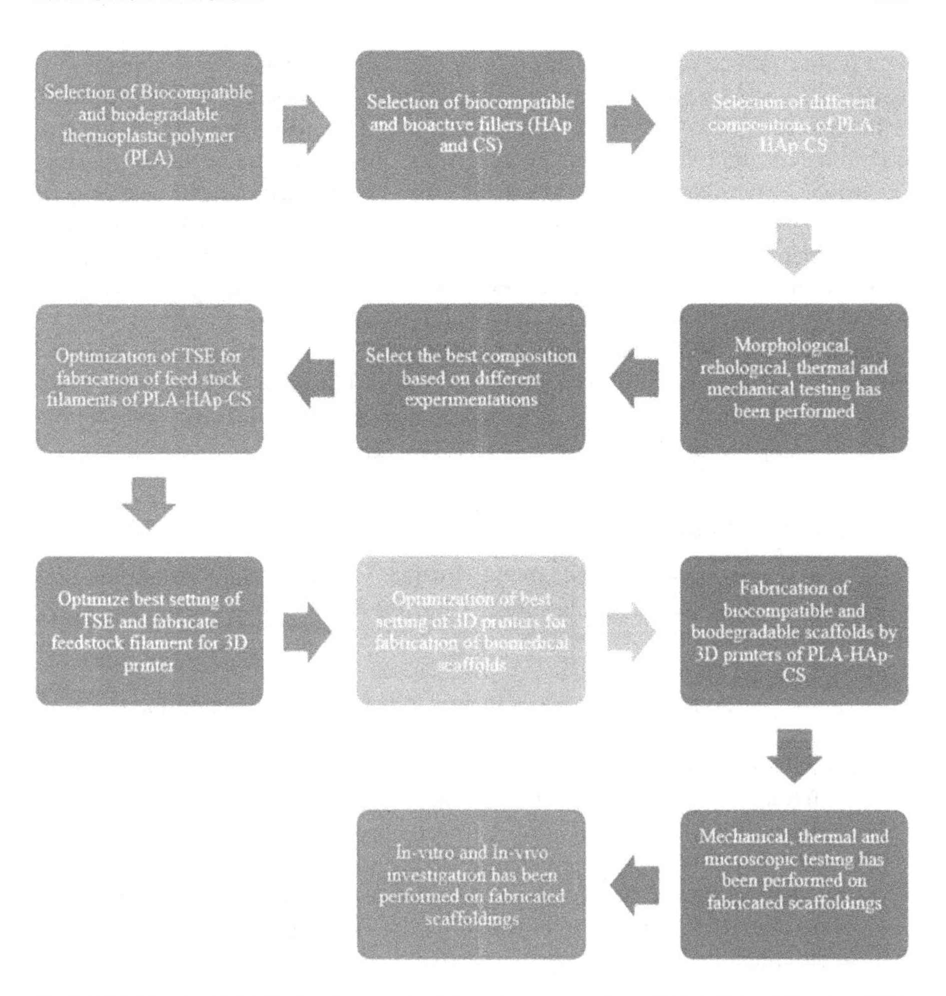

FIGURE 8.7 Scaffold fabrication methodology for processing of PLA-HAp-CS using TSE.

and biocompatible and bioactive fillers (PLA-HAp-CS) using TSE. The research work methodology of Ranjan et al. is described in Figure 8.7 (Ranjan et al., 2019c).

Based on previous research work, Ranjan et al. (2019) have worked on the fabrication of biocompatible, bioactive, and biodegradable scaffoldings by using the PLA thermoplastic polymer. In the first stage, the PLA thermoplastic polymer has been mixed with HAp and CS bioactive fillers as per literature study and pilot experimentations. Different experimentations (rheological, mechanical, thermal, dimensional, and flow-continuity) have been performed for the selection of the best composition. Finally, PLA 91% - HAp 8% - CS 1% (by weight %age) composition are out as a best composition according to different experimentations. In this research work for the fabrication of biocompatible scaffolds, the additive manufacturing (3D-printing) process has been used. Thermoplastic polymer–based feedstock filaments are used as input material in a 3D printer. In this research work, the best compositions (PLA 91% - HAp 8% - CS 1%) are used as an input material of TSE for fabrication of 3D-printer feedstock filament. For the fabrication of feedstock filament the optimization process has been performed for the best manufacturing setting. According to the Taguchi L9 (3*3) orthogonal array, nine different feedstock filaments have been prepared and after that tensile, dimensional, and microscopic analysis have been performed and the best setting of the TSE has been optimized. The best setting of the TSE feedstock filament for the 3D printer has been prepared. After fabrication of the feedstock filament, the 3D printer has been optimized according to the tensile (ASTM D638) and flexural (ASTM D790) samples that were manufactured based on the Taguchi L9 (3*3) orthogonal array. After that, biocompatible, bioactive, and biodegradable scaffoldings are printed on a 3D printer. Lastly, an in-vitro and in-vivo study have been performed on fabricated scaffoldings of the PLA-HAp-CS for confirmation of the biocompatibility test.

8.6 CONCLUSIONS

For this research work, "Twin-Screw Extrusion for Processing Thermoplastics in Biomedical Scaffolding Applications," the following observations have been concluded:

- Thermoplastic polymer is one of the most demanding materials due to its superior properties and ease of processing. The biomedical area is one of the most demanding areas of biocompatible and biodegradable polymers for fabrication of biomedical scaffoldings.
- According to the literature review, it has been concluded that a lot of research work has been performed on the processing of thermoplastic polymers, but much less has been reported on the reinforcement of thermoplastics with bioactive fillers for the fabrication of biomedical scaffoldings.

ACKNOWLEDGEMENT

The authors are thankful to the University Centre for Research and Development, Chandigarh University for their technical support.

REFERENCES

Aid, S., Eddhahak, A., Ortega, Z., Froelich, D., & Tcharkhtchi, A. (2017). Experimental study of the miscibility of ABS/PC polymer blends and investigation of the processing effect. *Journal of Applied Polymer Science*, *134*(25). 10.1002/app.44975

Amanat, N., James, N. L., & McKenzie, D. R. (2010). Welding methods for joining thermoplastic polymers for the hermetic enclosure of medical devices. *Medical Engineering and Physics*, *32*(7), 690–699. 10.1016/j.medengphy.2010.04.011

Asadi, N., Del Bakhshayesh, A. R., Davaran, S., & Akbarzadeh, A. (2020). Common biocompatible polymeric materials for tissue engineering and regenerative medicine. *Materials Chemistry and Physics*, *242*. 10.1016/j.matchemphys.2019.122528

Batoo, K. M., Kumar, G., Yang, Y., Al-Douri, Y., Singh, M., Jotania, R. B., & Imran, A. (2017). Structural, morphological and electrical properties of Cd2+doped MgFe2-xO4 ferrite nanoparticles. *Journal of Alloys and Compounds*, *726*, 179–186. 10.1016/j.jallcom.2017.07.237

Ben Difallah, B., Kharrat, M., Dammak, M., & Monteil, G. (2012). Mechanical and tribological response of ABS polymer matrix filled with graphite powder. *Materials & Design*, *34*, 782–787. 10.1016/j.matdes.2011.07.001

Braun, D. (2001). PVC? origin, growth, and future. *Journal of Vinyl and Additive Technology*, *7*(4), 168–176. 10.1002/vnl.10288

Canel, T., Kaya, A. U., & Çelik, B. (2012). Parameter optimization of nanosecond laser for microdrilling on PVC by Taguchi method. *Optics & Laser Technology*, *44*(8), 2347–2353. 10.1016/j.optlastec.2012.04.023

Crespo, J. E., Sánchez, L., García, D., & López, J. (2008). Study of the Mechanical and Morphological Properties of Plasticized PVC Composites Containing Rice Husk Fillers. *Journal of Reinforced Plastics and Composites*, *27*(3), 229–243. 10.1177/0731684407079479

Deshmukh, K., Houkan, M. T., AlMaadeed, M. A. A., & Sadasivunid, K. K. (2019). Introduction to 3D and 4D printing technology: State of the art and recent trends. In *3D and 4D Printing of Polymer Nanocomposite Materials: Processes, Applications, and Challenges*. Elsevier Inc. 10.1016/B978-0-12-816805-9.00001-6

Falahati, M., Ahmadvand, P., Safaee, S., Chang, Y. C., Lyu, Z., Chen, R., Li, L., & Lin, Y. (2020). Smart polymers and nanocomposites for 3D and 4D printing. *Materials Today*, *xxx*(xx). 10.1016/j.mattod.2020.06.001

Franzen, V., Kwiatkowski, L., Martins, P. A. F., & Tekkaya, A. E. (2009). Single point incremental forming of PVC. *Journal of Materials Processing Technology*, *209*(1), 462–469. 10.1016/j.jmatprotec.2008.02.013

Gatenholm, P., Bertilsson, H., & Mathiasson, A. (1993). The effect of chemical composition of interphase on dispersion of cellulose fibers in polymers. I. PVC-coated cellulose in polystyrene. *Journal of Applied Polymer Science*, *49*(2), 197–208. 10.1002/app.1993.070490202

Ha, C.-S., Kim, Y., Lee, W.-K., Cho, W.-J., & Kim, Y. (1998). Fracture toughness and properties of plasticized PVC and thermoplastic polyurethane blends. *Polymer*, *39*(20), 4765–4772. 10.1016/S0032-3861(97)10326-3

Hage, E., Ferreira, L. A. S., Manrich, S., & Pessan, L. A. (1999). Crystallization behavior of PBT/ABS polymer blends. *Journal of Applied Polymer Science*, *71*(3), 423–430. 10.1002/(SICI)1097-4628(19990118)71:3<423::AID-APP8>3.0.CO;2-0

Haghshenas, M. (2016). Metal–Matrix Composites. In *Reference Module in Materials Science and Materials Engineering*. Elsevier. 10.1016/B978-0-12-803581-8.03950-3

Han, J., Qi, J., Du, J., & Zhang, G. (2020). Fabrication of chitosan hydrogel incorporated with Ti-doped hydroxyapatite for efficient healing and care of joint wound. *Materials Letters*, *278*, 128415. 10.1016/j.matlet.2020.128415

Huneault, M. A., Carreau, P. J., Lafleur, P. G., & Gupta, V. P. (1992). Extrudate swelling and viscoelasticity of rigid PVC compounds. *Journal of Vinyl and Additive Technology*, *14*(4), 175–181. 10.1002/vnl.730140403

Kim, Y., Cho, W.-J., & Ha, C.-S. (1999). Dynamic mechanical and morphological studies on the compatibility of plasticized PVC/thermoplastic polyurethane blends. *Journal of Applied Polymer Science*, *71*(3), 415–422. 10.1002/(SICI)1097-4628(19990118)71:3<415::AID-APP7>3.0.CO;2-Z

Kumar, R., Chohan, J. S., Goyal, R., & Chauhan, P. (2020). Impact of process parameters of resistance spot welding on mechanical properties and micro hardness of stainless steel 304 weldments. *International Journal of Structural Integrity*, *12*(3), 366–377. 10.1108/IJSI-03-2020-0031

Lee, J. K., & Han, C. D. (2000). Evolution of polymer blend morphology during compounding in a twin-screw extruder. *Polymer*, *41*(5), 1799–1815. 10.1016/S0032-3861(99)00325-0

Lertwimolnun, W., & Vergnes, B. (2006). Effect of processing conditions on the formation of polypropylene/organoclay nanocomposites in a twin screw extruder. *Polymer Engineering & Science*, *46*(3), 314–323. 10.1002/pen.20458

Meijer, H. E. H., & Elemans, P. H. M. (1988). The modeling of continuous mixers. Part I: The corotating twin-screw extruder. *Polymer Engineering and Science*, *28*(5), 275–290. 10.1002/pen.760280504

Mousa, A., & Heinrich, G. (2010). Thermoplastic Composites Based on Renewable Natural Resources: Unplasticized PVC/Olive Husk. *International Journal of Polymeric Materials*, *59*(11), 843–853. 10.1080/00914037.2010.504143

Nadagouda, M. N., Rastogi, V., & Ginn, M. (2020). A review on 3D printing techniques for medical applications. *Current Opinion in Chemical Engineering*, *28*, 152–157. 10.1016/j.coche.2020.05.007

Nazeer, M. A., Onder, O. C., Sevgili, I., Yilgor, E., Kavakli, I. H., & Yilgor, I. (2020). 3D printed poly(lactic acid) scaffolds modified with chitosan and hydroxyapatite for bone repair applications. *Materials Today Communications*, *25*, 101515. 10.1016/j.mtcomm.2020.101515

Ognedal, A. S., Clausen, A. H., Polanco-Loria, M., Benallal, A., Raka, B., & Hopperstad, O. S. (2012). Experimental and numerical study on the behaviour of PVC and HDPE in biaxial tension. *Mechanics of Materials*, *54*, 18–31. 10.1016/j.mechmat.2012.05.010

Pita, V. J. R. R., Sampaio, E. E. M., & Monteiro, E. E. C. (2002). Mechanical properties evaluation of PVC/plasticizers and PVC/thermoplastic polyurethane blends from extrusion processing. *Polymer Testing*, *21*(5), 545–550. 10.1016/S0142-9418(01)00122-2

Ramteke, D. D., Balakrishna, A., Kumar, V., & Swart, H. C. (2017). Luminescence dynamics and investigation of Judd-Ofelt intensity parameters of Sm3+ ion containing glasses. *Optical Materials*, *64*, 171–178. 10.1016/j.optmat.2016.12.009

Ranjan, N. (2021). Chitosan with PVC polymer for biomedical applications: A bibliometric analysis. *Materials Today: Proceedings*. 10.1016/j.matpr.2021.04.274

Ranjan, N., Singh, R., & Ahuja, I. (2019a). Investigations for In-house prepared Biocompatible Feed Stock Filament of Fused Deposition Modelling: A Process Capability study. *Journal of Mechanical Engineering*, *48*(1), 18–23. 10.3329/jme.v48i1.41090

Ranjan, N., Singh, R., & Ahuja, I. P. S. (2019b). Investigations on joining of orthopaedic scaffold with rapid tooling. *Proceedings of the Institution of Mechanical Engineers, Part H: Journal of Engineering in Medicine*, *233*(7), 754–760. 10.1177/0954411919852811

Ranjan, N., Singh, R., & Ahuja, I. P. S. (2019c). Material processing of PLA-HAp-CS-based thermoplastic composite through fused deposition modeling for biomedical applications. *Biomanufacturing*, 123–136. 10.1007/978-3-030-13951-3_6

Ranjan, N., Singh, R., & Ahuja, I. P. S. (2020a). Development of PLA-HAp-CS-based biocompatible functional prototype: A case study. *Journal of Thermoplastic Composite Materials*, *33*(3), 305–323. 10.1177/0892705718805531

Ranjan, N., Singh, R., & Ahuja, I. P. S. (2020b). Mechanical, Rheological and Thermal Investigations of Biocompatible Feedstock Filament Comprising of PVC, PP and HAp. *Proceedings of the National Academy of Sciences India Section A - Physical Sciences*. 10.1007/s40010-020-00664-2

Ranjan, N., Singh, R., & Ahuja, I. P. S. (2021). Mechanical, Rheological and Thermal Investigations of Biocompatible Feedstock Filament Comprising of PVC, PP and HAp. *Proceedings of the National Academy of Sciences, India Section A: Physical Sciences*, *91*(1), 159–168. 10.1007/s40010-020-00664-2

Ranjan, N., Singh, R., Ahuja, I. P. S., Kumar, R., Singh, D., Ramniwas, S., Verma, A. K., & Mittal, D. (2021). 3D printed scaffolds for tissue engineering applications: Mechanical, morphological, thermal, in-vitro and in-vivo investigations. *CIRP Journal of Manufacturing Science and Technology*, *32*, 205–216. 10.1016/j.cirpj.2021.01.002

Ranjan, N., Singh, R., Ahuja, I. P. S., Kumar, R., Singh, J., Verma, A. K., & Leekha, A. (2020). On 3D printed scaffolds for orthopedic tissue engineering applications. *SN Applied Sciences*, *2*(2), 8–15. 10.1007/s42452-020-1936-8

Ranjan, N., Singh, R., Ahuja, I. P. S., & Singh, J. (2018). Fabrication of pla-hap-cs based biocompatible and biodegradable feedstock filament using twin screw extrusion. *Additive Manufacturing of Emerging Materials*, 325–345. 10.1007/978-3-319-91713-9_11

Ranjan, N., Singh, R., & Ahuja, I. S. (2020c). Development of HAp Reinforced Biodegradable Porous Structure Through Polymer Deposition Technology for Tissue Engineering Applications. In *Encyclopedia of Renewable and Sustainable Materials*. Elsevier Ltd. 10.1016/b978-0-12-803581-8.11264-0

Ranjan, N., Singh, R., & Ahuja, I. S. (2020d). Preparation of Partial Denture With Nano HAp-PLA Composite Under Cryogenic Grinding Environment Using 3D Printing. In *Encyclopedia of Renewable and Sustainable Materials*. Elsevier Ltd. 10.1016/b978-0-12-803581-8.11240-8

Ranjan, N., Singh, R., Ahuja, I. S., & Singh, J. (2017). A Framework for Development of Biocompatible Feedstock Filament of Polymers by Reinforcement of Fillers for FDM. *International Journal of Multidisciplinary Sciences and Engineering*, *8*(2), 185–189.

Rayate, A., & Jain, P. K. (2018). A Review on 4D Printing Material Composites and Their Applications. *Materials Today: Proceedings*, *5*(9), 20474–20484. 10.1016/j.matpr.2018.06.424

Sathiyavimal, S., Vasantharaj, S., LewisOscar, F., Selvaraj, R., Brindhadevi, K., & Pugazhendhi, A. (2020). Natural organic and inorganic–hydroxyapatite biopolymer composite for biomedical applications. *Progress in Organic Coatings*, *147*(June), 105858. 10.1016/j.porgcoat.2020.105858

Shunmugam, M. S., & Kanthababu, M. (2018). Advances in Additive Manufacturing and Joining. In *Editors Proceedings of AIMTDR*. http://www.springer.com/series/15734

Silva, J. V. L., & Rezende, R. A. (2013). Additive Manufacturing and its future impact in logistics. *IFAC Proceedings Volumes*, *46*(24), 277–282. 10.3182/20130911-3-BR-3021.00126

Singh, G., Pruncu, C. I., Gupta, M. K., Mia, M., Khan, A. M., Jamil, M., Pimenov, D. Y., Sen, B., & Sharma, V. S. (2019). Investigations of machining characteristics in the upgraded MQL-assisted turning of pure titanium alloys using evolutionary algorithms. *Materials*, *12*(6). 10.3390/ma12060999

Singh, J., Ranjan, N., Singh, R., & Ahuja, I. P. S. (2019). Multifactor Optimization for Development of Biocompatible and Biodegradable Feed Stock Filament of Fused Deposition Modeling. *Journal of The Institution of Engineers (India): Series E*, *100*(2), 205–216. 10.1007/s40034-019-00149-x

Singh, R., Kumar, R., Pawanpreet, Singh, M., & Singh, J. (2019). On mechanical, thermal and morphological investigations of almond skin powder-reinforced polylactic acid feedstock filament. *Journal of Thermoplastic Composite Materials*. 10.1177/0892705719886010

Singh, Rupinder, & Ranjan, N. (2018). Experimental investigations for preparation of biocompatible feedstock filament of fused deposition modeling (FDM) using twin screw extrusion process. *Journal of Thermoplastic Composite Materials*, *31*(11), 1455–1469. 10.1177/0892705717738297

Singh, Rupinder, Singh, R., Dureja, J. S., Farina, I., & Fabbrocino, F. (2017). Investigations for dimensional accuracy of Al alloy/Al-MMC developed by combining stir casting and ABS replica based investment casting. *Composites Part B: Engineering*, *115*, 203–208. 10.1016/j.compositesb.2016.10.008

Singh, S., Singh, N., Gupta, M., Prakash, C., & Singh, R. (2019). Mechanical feasibility of ABS/HIPS-based multi-material structures primed by low-cost polymer printer. *Rapid Prototyping Journal*, *25*(1), 152–161. 10.1108/RPJ-01-2018-0028

Sudeepan, J., Kumar, K., Barman, T. K., & Sahoo, P. (2014). Study of Friction and Wear of ABS/Zno Polymer Composite Using Taguchi Technique. *Procedia Materials Science*, *6*, 391–400. 10.1016/j.mspro.2014.07.050

Varma, M. V., Kandasubramanian, B., & Ibrahim, S. M. (2020). 3D printed scaffolds for biomedical applications. *Materials Chemistry and Physics*, *255*, 123642. 10.1016/j.matchemphys.2020.123642

Xu, S., Thomson, J. P., & Sahoo, M. (2004). A Review on Stress Relaxation and Bolt Load Retention of Magnesium Alloys for Automotive Applications. *Canadian Metallurgical Quarterly*, *43*(4), 489–506. 10.1179/cmq.2004.43.4.489

Zhu, D. Y., Cao, G. S., Qiu, W. L., Rong, M. Z., & Zhang, M. Q. (2015). Self-healing polyvinyl chloride (PVC) based on microencapsulated nucleophilic thiol-click chemistry. *Polymer*, *69*, 1–9. 10.1016/j.polymer.2015.05.052

Zhu, J., Tang, D., Lu, Z., Xin, Z., Song, J., Meng, J., Lu, J. R., Li, Z., & Li, J. (2020). Ultrafast bone-like apatite formation on highly porous poly(L-lactic acid)-hydroxyapatite fibres. *Materials Science and Engineering C*, *116*(May), 111168. 10.1016/j.msec.2020.111168

9 Case Study for the Development of a Hybrid Composite Structure of Thermosetting and Thermoplastics

Sanjeev Kumar, Rupinder Singh, Amrinder Pal Singh, and Yang Wei

9.1 INTRODUCTION

Plastic waste is increasing at a very fast rate, which will lead to its abundance in the future (Laria et al., 2020). The disposal of plastic waste has been done by burning and dumping, but these methods have led to toxic gases in the air and unavailability of land, respectively (Kumar et al., 2019). Although the recycling of thermoplastics has been done successfully, recycling of thermosetting plastics is still a challenge using conventional techniques. One of the most commonly used thermosetting plastics is bakelite (BAK), with properties such as high hardness, strength, and rigidity and contributing vastly to the plastic waste (Singh et al., 2019b). Low-density polyethylene (LPDE) has been used mostly as food packaging materials, sheets, and films. Properties such as good flexibility, transparency, fluidity, and glossy look (Salmah et al., 2013) with ability to adapt to high temperatures has made LDPE one of the most commonly used thermoplastics (Khattab et al., 2013).

A number of thermoplastics have been used as substrates for micro-strip patch antennas (MPA), providing lightweight, low cost, and robust design to MPA (Sarmah et al., 2010). Therefore, the recycled thermoplastics can be used as substrates for antenna applications, resulting in more valuable management of solid waste (bankey & Kumar, 2015). An antenna can be defined as a transmitter or receiver used for transmitting or receiving the signals from one system to another within a specific range wirelessly for a number of engineering applications (Sakib et al., 2020). MPA has shown a great application for

DOI: 10.1201/9781003184164-9

antenna applications due to its compactness, lightweight, simple profile, and planar design that can be fabricated easily using 3D printing (Khan et al., 2018). An MPA consists of a non-conductive substrate (with dielectric properties), a conductive radiating patch, and a conductive ground part. The radiating patch is printed on the top side and the ground part is printed on the bottom of the substrate (Salai Thillai & Ganesh Babu, 2018). The antenna has been used by different feeding techniques. Based on the different applications and the use of MPA, the patches can be different shapes, such as square, circular, rectangular, elliptical, or triangular (bankey & Kumar, 2015). The substrate also provides the mechanical strength to the MPA while allowing the wave to propagate through it (Kaur & Goyal, 2016). The ε_r determines the physical parameters of antennas and $\tan\delta$ defines the loss in a dielectric medium (Rida et al., 2007). Duroid® 5880, FR4, HK 04 J, Polyguide, and RF- 60 A are some of the most commonly used substrates for fabricating the MPA (Constantine A. Balanis, 2016). The most preferred feeding techniques are coaxial probe feed, microstrip line, aperture coupling, and proximity coupling (Singh & Tripathi, 2011). Providing a number of advantages, a patch antenna also has some disadvantages; low bandwidth is one of them (Dey & Mittra, 1996). In this research, 5% BAK has been added to LDPE for fabricating a substrate with a larger bandwidth.

3D printing has been a great concern for industries and researchers due to its ability to fabricate complicated parts at a low cost in a very short amount of time. 3D printing has been extensively studied for the design and manufacturing of mechanical structures (Ifwat et al., 2015). The materials such as polymers, metals, ceramics, concrete, or even biological tissues have been printed successfully using the different 3D-printing techniques (Liang et al., 2015). Ease of fabricating compact shapes on 3D printing has shown to be quite useful in fabricating the miniaturised MPAs (Ifwat et al., 2015). One of the most used 3D-printing techniques is FDM, offering flexibility in design (Kumar et al., 2020).

This work is an extension of previous reported studies on a comparison of thermoplastic and thermoplastic-thermosetting composite-based 3D-printed patch antenna (Singh et al., 2022). This study was only limited to the simulation of the LDPE and LDPE-5%BAK patch antennas and comparing their simulated parameters, such as S_{11} parameters and gain at 2.45 GHz. But, in this work, LDPE and LDPE-5%BAK composite antennas have been simulated and their parameters, such as S_{11} parameters, gain, and VSWR at 3 GHZ, have been compared and both antennas have also been designed and fabricated to validate the simulated results. Finally, the composites of LDPE and BAK have led to a useful product in the form of MPA with improved antenna properties at 3 GHz.

9.2 LITERATURE SURVEY

In previous studies, it has been reported that thermoplastics such as acrylonitrile butadiene styrene (ABS) have shown their application in antenna design (Abdul Malek et al., 2017). In a different study, researchers have used FDM for fabricating and characterising a microwave MPA resonating at 7.5 GHz using an ABS as 3D printed substrate (Liang et al., 2015). LDPE with some reinforcements has been used as a substrate for MPA (Sarmah et al., 2010). It was also observed in different research that there was not any monoatomic variation of ε_r of LDPE by thermal ageing at 90°C for 2,500 hours. Design parameters for a rectangular patch antenna using an inset feed has been discussed by the researcher in their work defining the dependency of return loss, VSWR, directivity, gain, and bandwidth on notch width (Rahman et al., 2020). In another research, the author has discussed the effect of patch dimensions and dielectric properties of substrates on radiation parametres of bandwidth and beamwidth (Paul & Sultan, 2013). A MPA with a rectangular patch resonating at 3 GHz has been simulated and fabricated for applications in weather radar and communication satellites. The simulation has been done using the CST software (Mynuddin, 2020). A change in performance parameters has also been observed by changing the substrate material (Epoxy_kevlar_xy with dielectric constant 3.6 and FR4_epoxy with dielectric constant 4.4). It was observed that a decrease in ε_r increased the bandwidth and shows better efficiency (Kaur & Goyal, 2016). In another study, a comparison of RF and microwave performance has been done by using poly methyl methacrylate (PMMA) and poly lactic acid (PLA) as substrate material with FR4 as substrate material for the MPA (Manab et al., 2018).

Researchers have successfully reinforceed 10% BAK in ABS concerning the MFI characteristics. The feedstock filament shows better tensile properties and has been extruded using twin-screw extrusion (Singh et al., 2019b). Further, the ABS-10%BAK composite has been successfully printed using FDM with a different infill ratio and printing speeds (Singh et al., 2019a). LDPE has been studied as outdoor structures such as window frames, pipes, and automobiles (Ovalı, 2020). It has been observed that thermosetting plastics reinforced in thermoplastic can be 3D printed on FDM.

9.3 RESEARCH GAP

In order to ascertain the research gap for thermoplastics and thermoplastic-thermosetting composite-based 3D-printed patch antenna, bibliographic analysis was performed by using the Web of Science database. Initially, the search was performed with the keyword patch antenna and 13,136 results were obtained (for the selected period 1999–2021). Out of these results, the most recent 1,000 results have been shortlisted for further analysis using VOS Viewer open-source software. By selecting a minimum number of occurrences of the term as "5" out of 10,698 terms, 697 met the threshold. For each of the 697 terms, a relevance score was calculated and, based upon this score, 60% of the most relevant terms (418) were selected for final analysis. By using these 418 terms, a networking diagram asa bibliographic analysis has been prepared (Figure 9.1).

Similarly, another search was performed for the keyword "3D printed patch antenna" and 76 results were obtained. Figures 9.2 and 9.3, respectively, show bibliographic analysis and gap analysis for the selected keyword. Another search

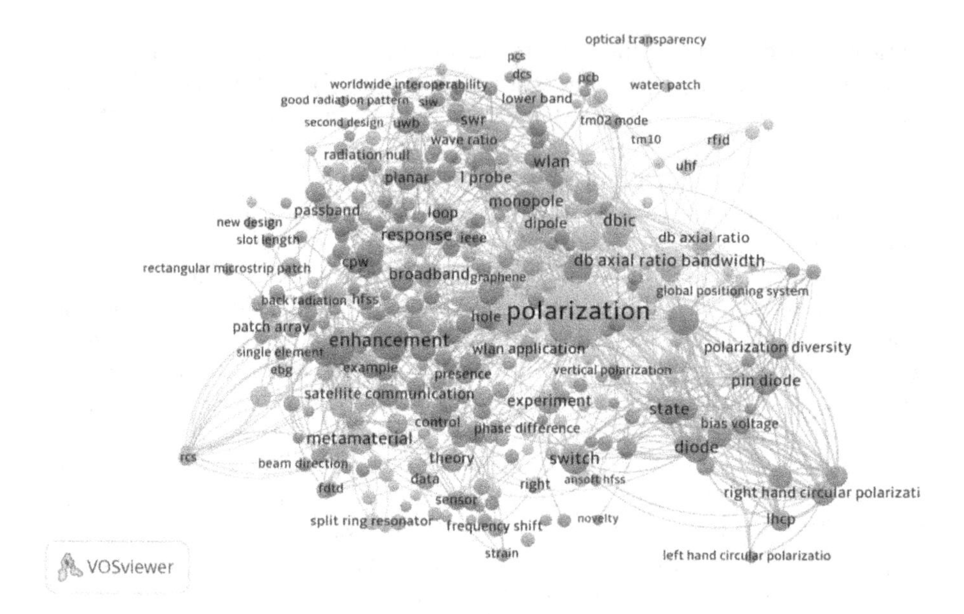

FIGURE 9.1 Bibliographic analysis based upon keyword patch antenna (Based on VOS Viewer software).

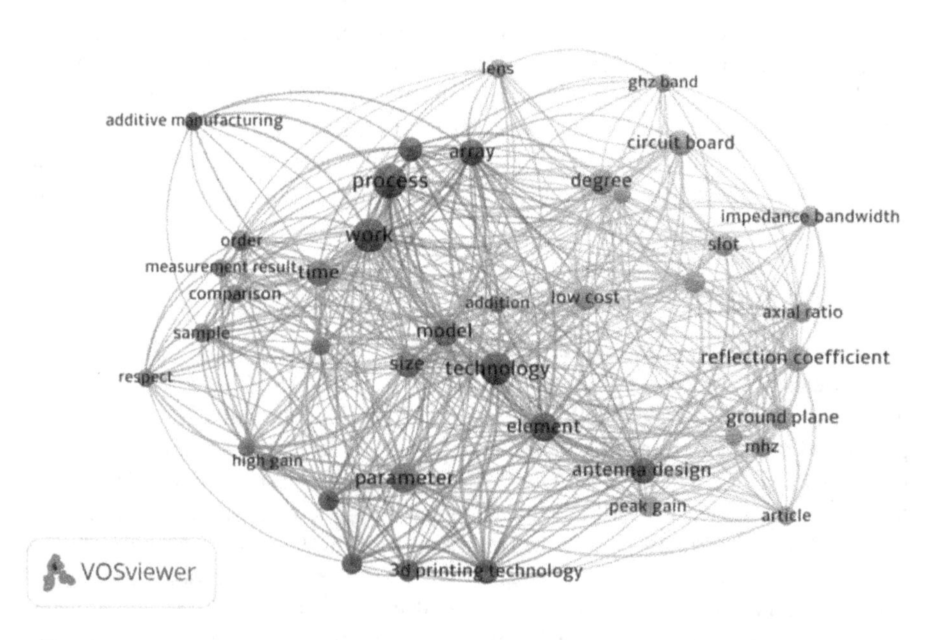

FIGURE 9.2 Bibliographic analysis based upon keyword 3D-printed patch antenna (Based on VOS viewer software).

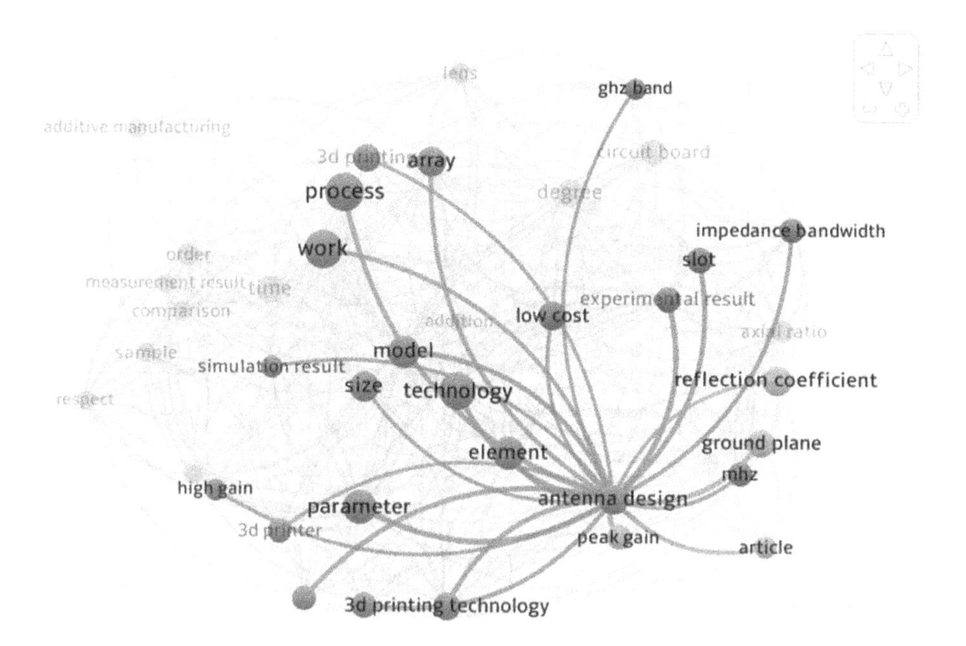

FIGURE 9.3 Gap analysis based upon keyword 3D-printed patch antenna (Based on VOS viewer software).

was performed for the keyword "thermoplastic patch antenna" and 12 results were obtained. Finally, for "thermosetting patch antenna" as a keyword, no result was obtained in the Web of Science core collection. The literature review reveals that significant studies have been reported on patch antennas and also some work has been reported on 3D-printed patch antennas. But limited studies have been reported on thermoplastic patch antennas and no work has been reported on thermoplastic-thermosetting composite-based 3D-printed substrate for patch antennas and their mechanical, morphological, rheological, and radiofrequency (RF) characterization for sensing applications. Further, the use of thermoplastic and thermosetting-based substrates in patch antennas may be treated as a novel way of secondary recycling, resulting in development of a high-end value-added product. So, in this study, LDPE and LDPE-5% BAK composite-based substrate were printed (for secondary recycling) on FDM for preparation of patch antennas (resonating frequency 3 GHz).

9.4 METHODOLOGY AND EXPERIMENTATION

This work is an extension of our previous work (Singh et al., n.d.) in which the study was only limited till the simulation of LDPE and LDPE-5%BAK-based MPA for resonating at 2.45 GHz. In this work, patch antennas have been designed, simulated, and fabricated for a resonating frequency of 3 GHz for applications such as weather radar and communication satellites. The methodology adopted for this work has been shown in Figure 9.4.

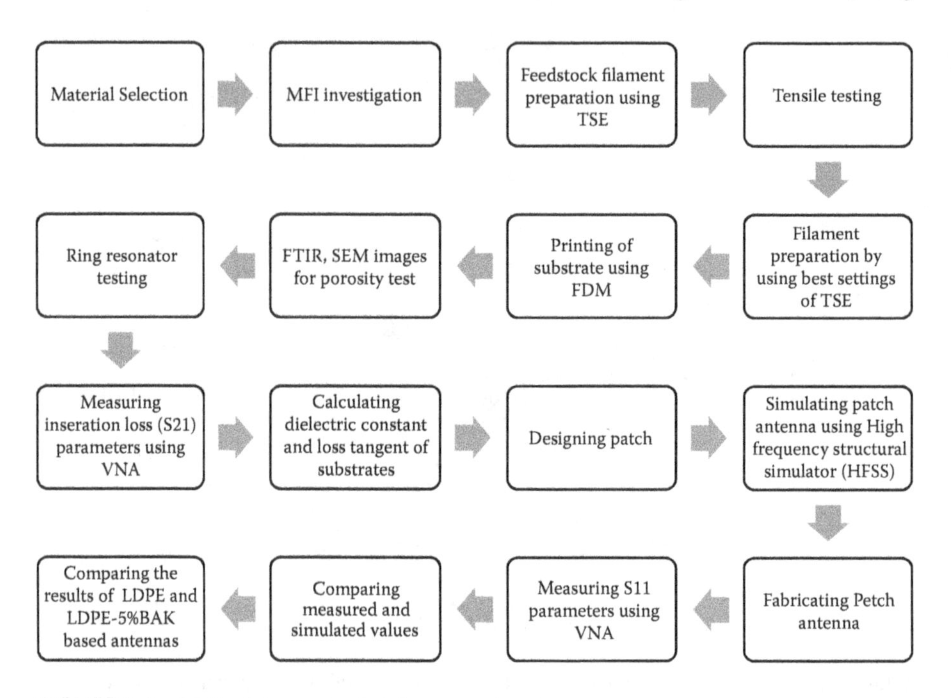

FIGURE 9.4 Methodology adopted for present study.

TABLE 9.1
MFI observations

S. No.	LDPE (Wt.%)	BAK (Wt.%)	MFI in g/(10 min)
1	100%	0%	12.15 ± 0.05
2	97.5%	2.5%	11.45 ± 0.05
3	95%	5%	9.90 ± 0.06

The melt flow index values for raw LDPE and LDPE with 5% bakelite has been measured as 12.15 ± 0.05 g/10 min and 9.90 ± 0.06 g/10 min, respectively, as shown in Table 9.1.

Wire samples of LDPE-5%BAK have been extruded on the twin-screw extruder at different settings (such as 180,185,180 screw temperature (°C); 90,100,110 rpm; and 5,8,11 load) using the L9 array for optimizing the peak and break strength of the wire to be extruded. These samples were then tested on a universal testing machine. The observed values on the UTM have been shown in Table 9.2.

The substrates of size 50 × 40 × 0.65 mm have been printed using the source FDM (Make: ultimaker^{2+}) shown in Figure 9.5. The nozzle and bed temperature were kept at 230°C and 60°C, respectively, while printing both substrates.

TABLE 9.2
Observed mechanical properties

Sample	Peak Elongation (mm)	Peak Strength (MPa)	Break Elongation (mm)	Break Strength (MPa)
LDPE	5.4 ± 0.06	11.9 ± 0.08	7.1 ± 0.04	10.7 ± 0.09
C1	3.9 ± 0.09	11.1 ± 0.09	6.9 ± 0.03	10.0 ± 0.07
C2	3.3 ± 0.06	9.3 ± 0.08	4. ± 0.02	8.4 ± 0.04
C3	4.8 ± 0.03	16.0 ± 0.08	4.8 ± 0.03	14.4 ± 0.07
C4	6.7 ± 0.02	9.0 ± 0.09	7.7 ± 0.07	8.0 ± 0.09
C5	6.9 ± 0.03	15.3 ± 0.09	7.1 ± 0.04	13.8 ± 0.05
C6	5.2 ± 0.05	13.2 ± 0.09	5.4 ± 0.06	11.9 ± 0.06
C7	2.9 ± 0.04	12.2 ± 0.09	3.1 ± 0.05	11.0 ± 0.06
C8	3.7 ± 0.08	13.6 ± 0.08	3.9 ± 0.09	12.3 ± 0.01
C9	4.4 ± 0.01	11.2 ± 0.09	4.4 ± 0.01	10.1 ± 0.06

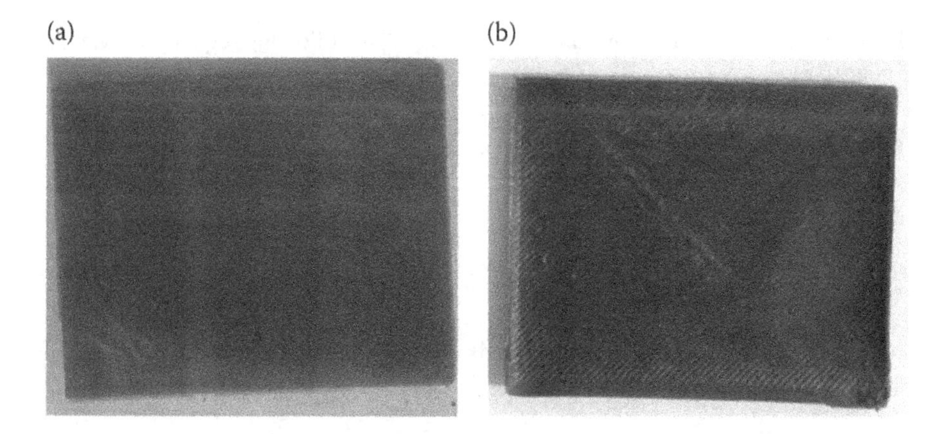

FIGURE 9.5 3D-printed substrates (a) LDPE and (b) LDPE-5%BAK.

9.5 RESULTS AND DISCUSSION

In previous work, it has been observed that the LDPE-5%BAK has shown better mechanical properties supported by porosity and SEM results. It was also observed that the LDPE-5%BAK substrate has shown better transmittance compared to the LDPE substrate (using attenuated total reflection (ATR)–based Fourier transformed infrared (FTIR) analysis).

9.5.1 Design of the Patch Antenna

The patch antenna has been designed by specifying the parameters such as the resonating frequency, patch length, patch width, and effective dielectric constant for each material. In this study, the patch antenna has been designed for a resonating frequency of 3 GHz with 0.65 mm substrate thickness. The ε_r for LDPE and LDPE-5%BAK was 2.282 and 2.0663, respectively, and $\tan\delta$ for LDPE and LDPE-5%BAK was 0.0045 and 0.0051, respectively, which have been calculated in a previous work (Singh et al., n.d.).

Using the value of ε_r dimensions of the patch and substrate has been calculated for each LDPE and LDPE-5% BAK.

The equation given below gives the width of the patch of antenna (Manab et al., 2018):-

$$W = \frac{c}{2f_r}\sqrt{\left(\frac{2}{\epsilon_r + 1}\right)} \tag{9.1}$$

where

W= width of the patch
f_r = resonant frequency
ϵ_r = dielectric constant

The effective dielectric constant is given by the following equation (Manab et al., 2018):-

$$\epsilon_{eff} = \left[\frac{\epsilon_r + 1}{2}\right] + \left[\left(\frac{\epsilon_r - 1}{2}\right)\left(1 + 12\frac{h}{W}\right)^{-0.5}\right] \tag{9.2}$$

where

h = thickness of the substrate

The value of ΔL, the fring factor, needs to be calculated to find the length of the patch (Manab et al., 2018):-

$$\Delta L = 0.42h\left[\frac{(\epsilon_{eff} + 0.3)\left(\frac{w}{h} + 0.264\right)}{(\epsilon_{eff} - 0.258)\left(\frac{w}{h} + 0.8\right)}\right] \tag{9.3}$$

The length of the patch is calculated as (Manab et al., 2018):

$$L = \frac{1}{2f_c\sqrt{(\epsilon_{eff}\,\epsilon_0\,\mu_0)}} - 2\Delta L \tag{9.4}$$

The minimum required length and width of the substrate are obtained by the following equations (bankey & Kumar, 2015):

$$W_s = W + 6h \tag{9.5}$$

TABLE 9.3
Dimensions of patch and substrates for both antennas at resonating frequency 3 GHz

Substrate	Length of Patch (mm)	Width of Patch (mm)	Length of Substrate (mm)	Width of Substrate (mm)
LDPE	32.81	39	36.71	42.9
LDPE + 5%BAK	34.45	40.35	38.35	44.25

$$L_s = L + 6h \quad (9.6)$$

Where W_s and L_s are the width and length of the substrate, respectively, and h is the thickness of the substrate.

Table 9.3 shows the calculated dimensions for length and width of the patch and substrate (using equations 9.1–9.6).

The length and width of the substrates have been taken as 40 mm and 50 mm for both antennas.

9.5.2 Simulation of Antennas Using HFSS

Further simulation for patch antennas was performed by using a high-frequency structure simulator (HFSS) 15.0.3 to simulate the patch antennas for a resonating frequency of 3 GHz (Figure 9.6). A coaxial feed has been used for the excitation for both antennas.

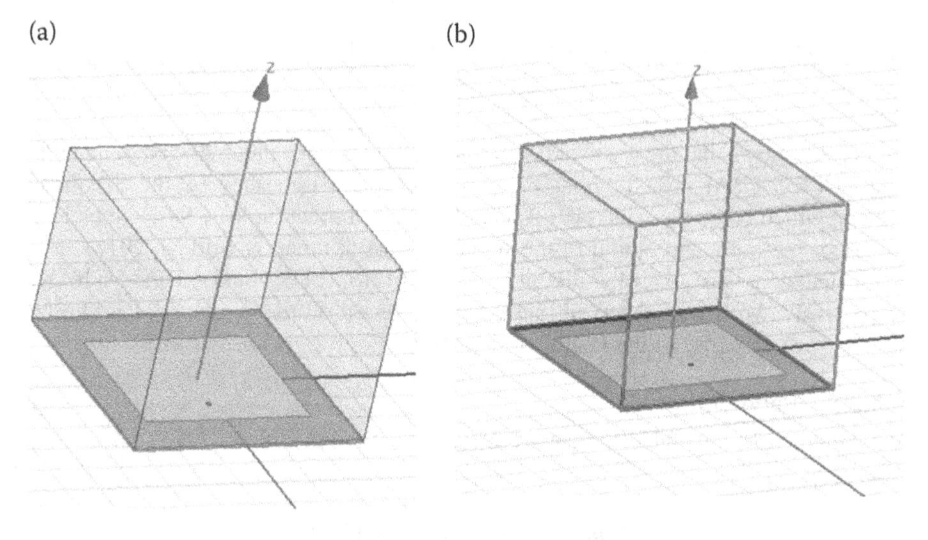

FIGURE 9.6 Design of MPA (a) LDPE, (b) LDPE-5%BAK.

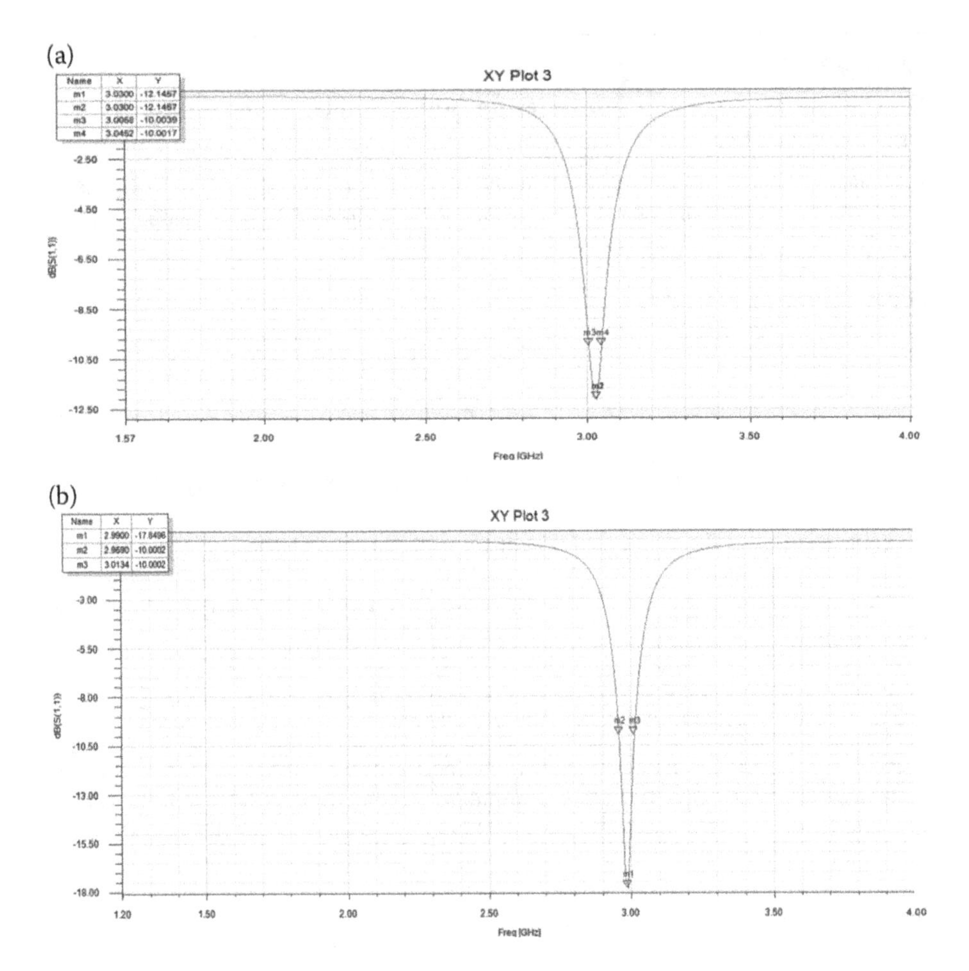

FIGURE 9.7 S_{11} Parameters at 3 GHz for (a) LDPE and (b) LDPE-5%BAK.

9.5.2.1 S_{11} Parameters

Figure 9.7 shows that the antenna is resonating at 3.03 GHz and 2.99 GHz for LDPE and LDPE-5%BAK-based MPA. Losses in MPA are described by the S_{11} parameters, which when described in dB, can be termed reflection loss. Reflections loss describes the input and output of the signal sources (Kumari & Sridevi, 2016). The S_{11} calculated was −12.1457 dB and −17.84 dB for LDPE and LDPE-5%BAK, respectively. So it was observed that the MPA based on the composite has shown low losses compared to the LDPE-based MPA.

9.5.2.2 Bandwidth

Bandwidth was calculated by using the formula (Paul & Sultan, 2013):

$$Bandwidth = \frac{f_1 - f_2}{f_1 * f_2} \times 100\%$$

The values of f_1 and f_2 were taken at –10 dB from Figure 9.7. The bandwidth for MPA of LDPE and LDPE-5%BAK was calculated as 1.30% and 1.82%, respectively, which shows the composite-based MPA has better bandwidth.

9.5.2.3 The Voltage Standing Wave Ratio (VSWR)

The ratio of maximum voltage to minimum voltage in a standing wave pattern is developed when power is reflected from the load. Therefore, VSWR defines the power delivered to the device and it is also a measure of impedence matching of the source and load. The purposed antennas have shown a VSWR of 2.1 and 1.49 for LDPE and LDPE-5%BAK, respectively, at 3 GHz, as shown in Figure 9.8.

9.5.2.4 Gain

The gain of an antenna (in a given direction) is defined as "the ratio of the intensity, in a given direction, to the radiation intensity that would be obtained if the power

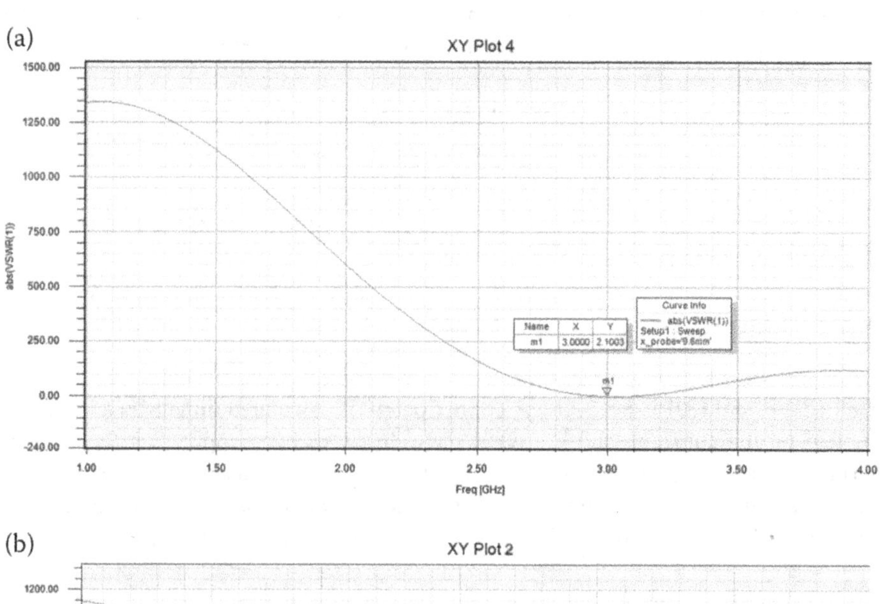

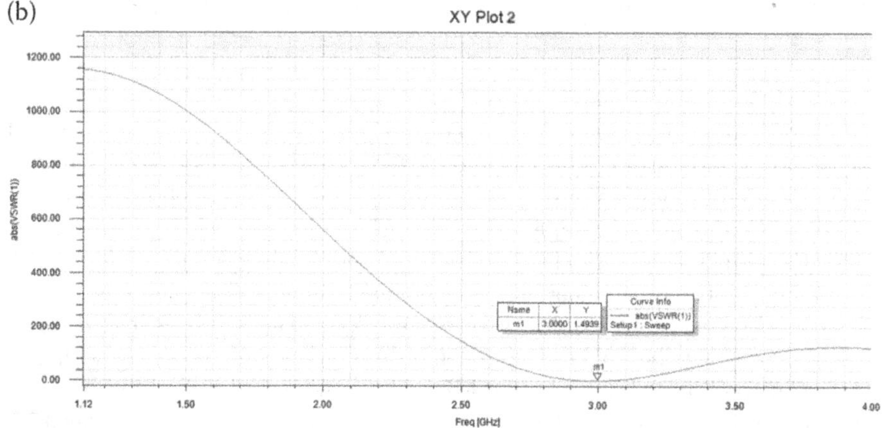

FIGURE 9.8 VSWR at 3 GHz (a) LDPE and (b) LDPE-5%BAK.

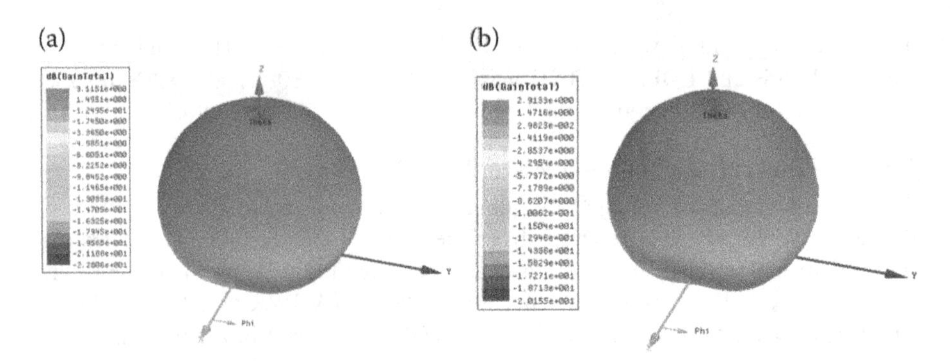

FIGURE 9.9 Gain at 3 GHz (a) LDPE and (b) LDPE-5%BAK.

accepted by the antenna were radiated isotropically. The radiation intensity corresponding to the isotropically radiated power is equal to the power accepted (input) by the antenna divided by 4 ." In equation form, this can be expressed as

$$\text{Gain} = 4\pi\frac{\text{Radiation intensity}}{\text{total input (accepted) power}} = 4\pi\frac{U(\theta, \phi)}{P_{in}}$$

The total gain was simulated as 3.115 dB and 2.913 dB for LDPE and LDPE-5% BAk, respectively, as shown in Figure 9.9.

9.5.3 Fabricating of a Patch Antenna

The substrate of size 40 * 50 * 0.65 mm was printed using ultimaker 2+ one for each LDPE and LDPE-5%BAK. A copper patch of calculated dimensions has been applied on the substrate and a ground plane has also been applied on the bottom surface of the substrate. The connector has been soldered as a coaxial probe feed, as shown in Figure 9.10. A copper tape of thickness 0.08 mm was used as a conductor for the patch and ground. An SMA connector was used as a coaxial feed.

9.5.4 VNA Testing

The fabricated antennas were then tested on the VNA. At first the calibration of the VNA was done using a kit available with a VNA setup. The S_{11} parameters ware measured using the VNA, as shown in Figure 9.11. It was observed that the LDPE was resonating at 3.09 GHz and LDPE-5%BAK was resonating at 2.99 GHz, which were very close to the simulated results shown in Figure 9.12, confirming the results obtained on the fabricated antenna. The measurements of S_{11} were –9.031 dB and 16.3867 dB for LDPE and LDPE-5%BAK-based MPA, which shows that the LDPE-5%BAK has been transmitting signals with low losses compared to the LDPE-based MPA, conforming the higher transmittance observed in the attenuated total reflection (ATR)–based Fourier transformed infrared (FTIR) analysis.

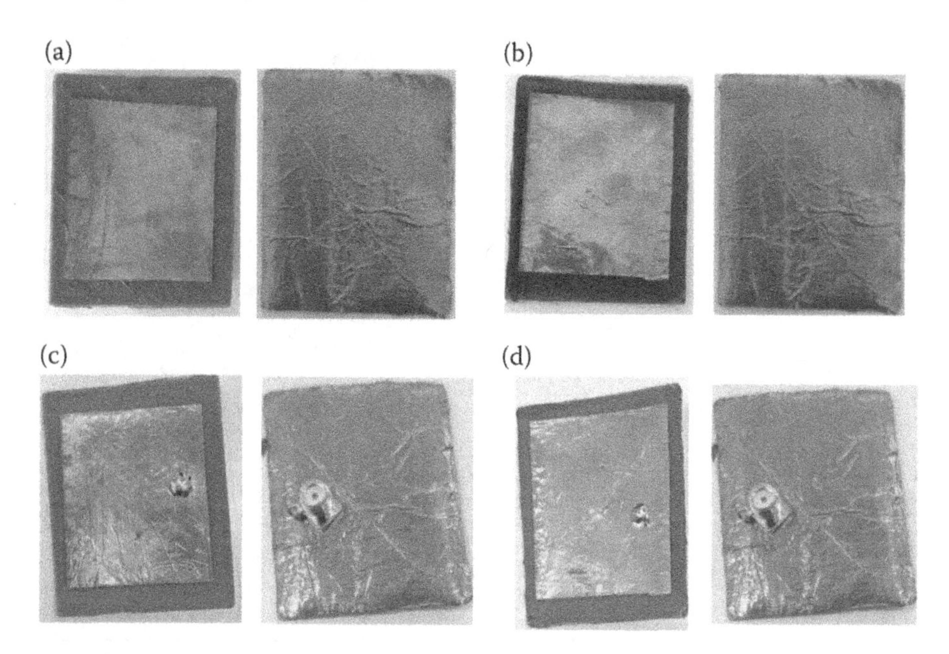

FIGURE 9.10 Fabrication of MPA; front and back view after applying patch and ground (a) LDPE, (b) LDPE-5%BAK; front and back view after soldering SMA connector (c) LDPE, (d) LDPE-5%.

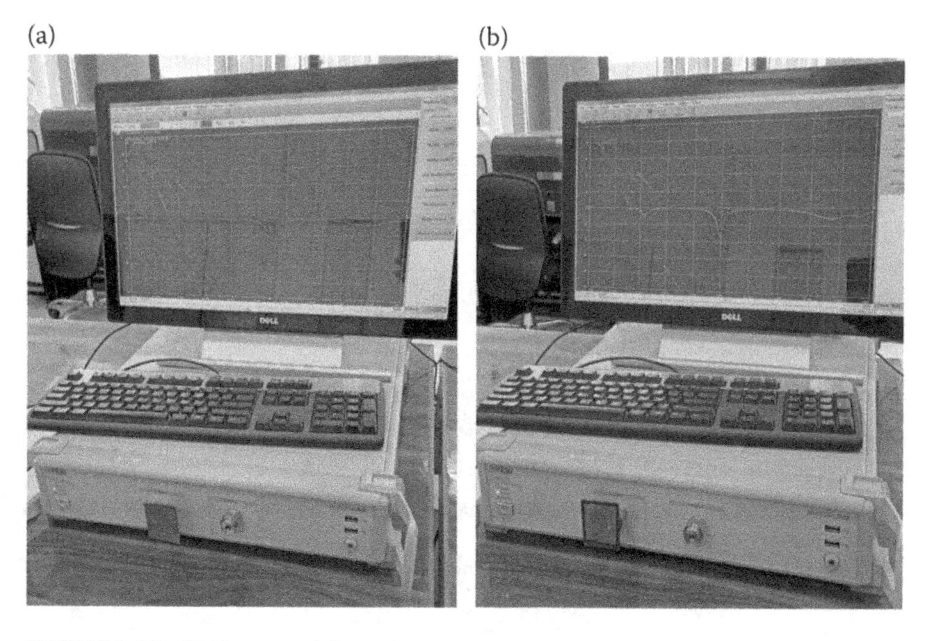

FIGURE 9.11 VNA setup while testing (a) LDPE; (b) LDPE-5%BAK.

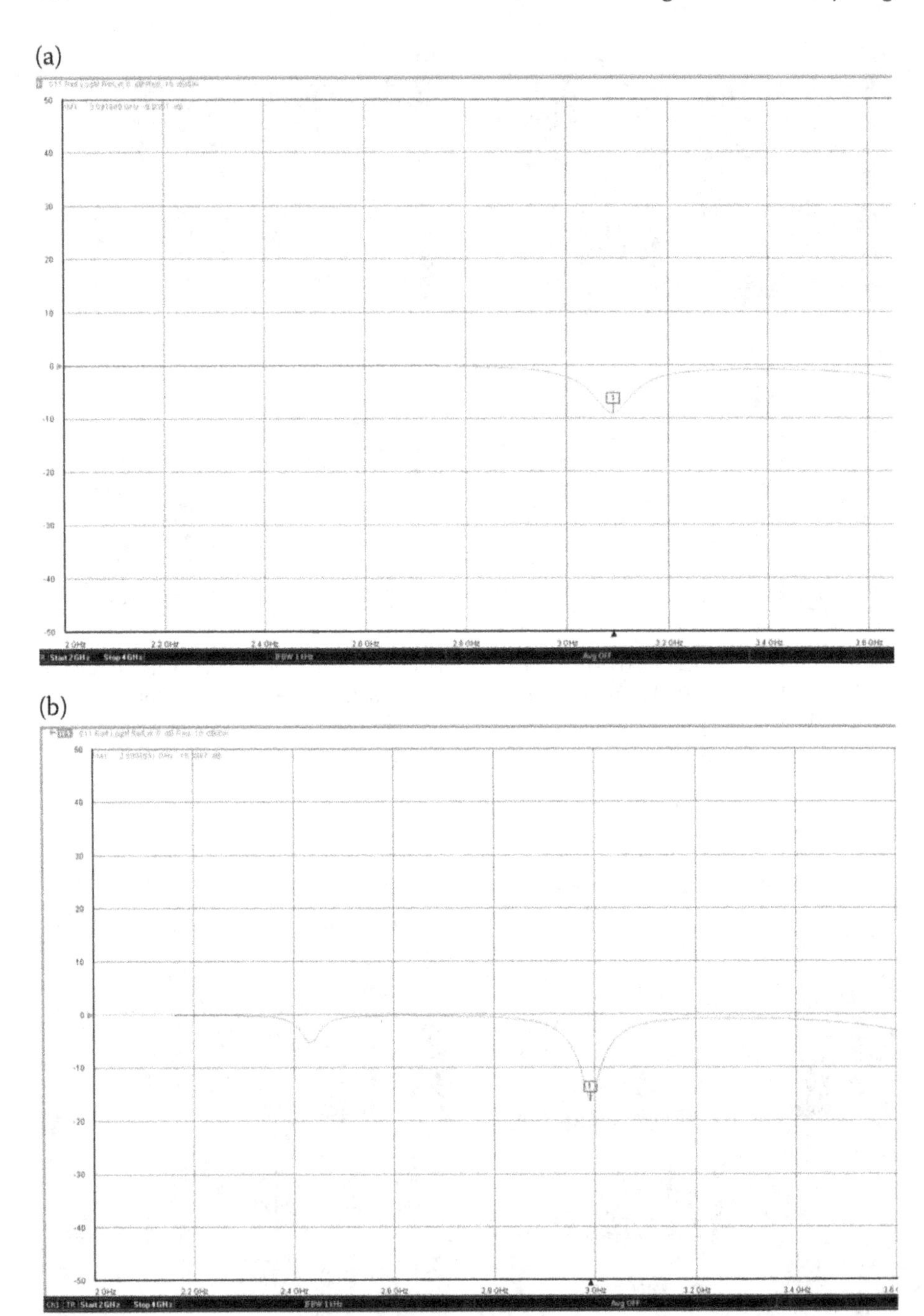

FIGURE 9.12 S_{11} parameters measured using VNA (a) LDPE and (b) LDPE-5%BAK.

9.6 CONCLUSIONS

Following are the findings of this study:

1. The antenna prepared from the thermoplastic and thermosetting waste has been working within the required range. Hence, it can be used as an option for converting the solid plastic waste in the form of some usable physical application (in this case, radar and satellite applications).
2. The MPA obtained from LDPE-5%BAK has shown a better transmittance, better bandwidth, and low losses compared to the LDPE-based MPA. Hence, this led to better waste management for thermosetting plastic waste.
3. It was observed that the reinforcement of 5% bakelite in LDPE has decreased its MFI from 12.15 ± 0.05 g/10 min to 9.90 ± 0.06 g/10 min, but both values lie in a range of printing on the FDM setup.

9.7 FUTURE SCOPE

Further studies may be performed by customizing other mechanical properties, such as modulus of toughness, Young's modulus, peak elongation, etc., as required by the antenna applications. Heat capacity of LDPE-5%BAK in comparison to LDPE can also be considered for application of sensors at elevated temperatures. The design of MPA based on LDPE-5%BAK can also be varied for increasing the bandwidth as it has already provided better bandwidth with a simple design.

ACKNOWLEDGEMENT

This research has been funded under NTU-PU collaborated project titled "Wearable 3D printed patch antenna." Most parts have been done at the manufacturing research lab, GNDEC, Ludhiana and National Institute of Technical Teachers Training and Research.

REFERENCES

Abdul Malek, N., A. M. Ramly, A. Sidek, & S. Y. Mohamad. (2017). Characterization of Acrylonitrile Butadiene Styrene for 3D Printed Patch Antenna. *Indonesian Journal of Electrical Engineering and Computer Science*, *6*(1), 116–123. 10.11591/ijeecs.v6.i1.pp116-123.

Bankey, V., & N. A. Kumar. (2015). Design and Performance Issues of Microstrip Antennas. *International Journal of Scientific and Engineering Research*, *6*(3), 1572–1580. 10.14299/ijser.2015.03.008.

C. A. Balanis. (2016). *Antenna Theory Analysis and Design*. 4th edition. John Wiley & Sons, Inc., Hoboken, New Jersey.

Dey, S., & R. Mittra. (1996). *A Compact Broadband Microstrip Antenna. Microwave and Optical Technology Letters*. Vol. *11*. 10.1002/(sici)1098-2760(19960420)11:6<295::aid-mop2>3.0.co;2-e.

Ifwat, M., M. Ghazali, E. Gutierrez, J. C. Myers, A. Kaur, B. Wright, & P. Chahal. (2015). Affordable 3D Printed Microwave Antennas, IEEE 65th Electronic Components and Technology Conference, San Diego, CA, USA, 240–246.

Kaur, G., & S. Goyal. (2016). To Study the Effect of Substrate Material for Microstrip Patch Antenna. *International Journal of Engineering Trends and Technology*, *36*(9), 490–493. 10.14445/22315381/ijett-v36p289.

Khan, I., Geetha, D., R. Gunjal, & R. Rashmitha. (2018). Review of MSP Antenna Design for Various Substrates. *SSRN Electronic Journal*, 1–4. 10.2139/ssrn.3230632.

Khattab, A., C. Liu, W. Chirdon, & C. Hebert. (2013). Composite Materials. 10.1177/0892705711432361.

Kumar, S., R. Singh, & A. Batish. (2019). On Investigation of Rheological, Mechanical and Morphological Characteristics of Waste Polymer-Based Feedstock Filament for 3D Printing Applications. 10.1177/0892705719856063.

Kumar, S., R. Singh, T. P. Singh, & A. Batish. (2020). Comparison of Mechanical and Morphological Properties of 3-D Printed Functional Prototypes: Multi and Hybrid Blended Thermoplastic Matrix, 1–16. 10.1177/0892705720925136.

Kumari, K. K., & Prof P. V. Sridevi. (2016). Design and Analysis of Microstrip Antenna Array Using CST Software, *6*(5), 5977–5982. 10.4010/2016.1449.

Laria, G., R. Gaggino, L. E. Peisino, & A. Cappelletti. (2020). Mechanical and Processing Properties of Recycled PET and LDPE-HDPE Composite Materials for Building Components. 10.1177/0892705720939141.

Liang, M., S. Member, C. Shemelya, E. Macdonald, R. Wicker, H. Xin, & Sr Member. (2015). 3D Printed Microwave Patch Antenna via Fused Deposition Method and Ultrasonic Wire Mesh Embedding Technique, *1225* (c): 10–13. 10.1109/LAWP.2015.2405054.

Manab, N. H., E. Baharudin, F. C. Seman, & A. Ismail. (2018). 2.45 GHz Patch Antenna Based on Thermoplastic Polymer Substrates. *RFM 2018 - 2018 IEEE International RF and Microwave Conference, Proceedings*, 93–96. 10.1109/RFM.2018.8846493.

Mynuddin, M. (2020). Design and Simulation of High Performance Rectangular Microstrip Patch Antenna Using CST Microwave Studio. *Global Scientific Journal*, *8*(8), 2225–2229.

Ovalı, S. (2020). Investigating the Effect of the Aging Process on LDPE Composites with UV Protective Additives, 1–19. 10.1177/0892705720941908.

Paul, L. C., & N. Sultan. (2013). Design, simulation and performance analysis of aline feed rectangular micro-strip patch antenna. *International Journal of Engineering Sciences & Emerging Technologies*, *4*(2): 117–126.

Rahman, Z., M. Mynuddin, & K. C. Debnath. (2020). The Significance of Notch Width on the Performance Parameters of Inset Feed Rectangular Microstrip Patch Antenna, *10*(1), 7–18. 10.5923/j.ijea.20201001.02.

Rida, A., L. Yang, R. Vyas, S. Bhattacharya, & M. M. Tentzeris. (2007). Design and Integration of Inkjet-Printed Paper-Based UHF Components for RFID and Ubiquitous Sensing Applications. In 2007 European Microwave Conference, pp. 724–727, October: 724–727.

Sakib, N., S. N. Ibrahim, M. I. Ibrahimy, M. S. Islam, & M. M. Hasan Mahfuz. (2020). Design of Microstrip Patch Antenna on Rubber Substrate with DGS for WBAN Applications. *2020 IEEE Region 10 Symposium, TENSYMP 2020*, no. August: 1050–1053. 10.1109/TENSYMP50017.2020.9230707.

Salai Thillai, T. J., & T. R. Ganesh Babu. (2018). Rectangular Microstrip Patch Antenna at ISM Band. *Proceedings of the 2nd International Conference on Computing Methodologies and Communication, ICCMC 2018*, no. Iccmc: 91–95. 10.1109/ICCMC.2018.8487877.

Salmah, H., A. Romisuhani, & H. Akmal. (2013). Properties of Low-Density Polyethylene/ Palm Kernel Shell Composites: Effect of Polyethylene Co-Acrylic Acid. 10.1177/0892705711417028.

Sarmah, D., J. R. Deka, & S. Bhattacharyya. (2010). Study of LDPE / TiO 2 and PS / TiO 2 Composites as Potential Substrates for Microstrip Patch Antennas, *39*(10): 2359–2365. 10.1007/s11664-010-1335-9.

Singh, I., & V. S. Tripathi. (2011). Micro Strip Patch Antenna and Its Applications: A Survey. *International Journal of Computer Applications in Technology*, *2*(5), 1595–1599.

Singh, R., S. Kumar, A. P. Singh, & Y. Wei. n.d. On comparison of recycled LDPE and LDPE–bakelite composite based 3D printed patch antenna, Journal of Materials: Design and Applications, IMech Part L, 2022, In-Press.

Singh, R., I. Singh, & R. Kumar. (2019a). Mechanical and Morphological Investigations of 3D Printed Recycled ABS Reinforced with Bakelite–SiC–Al2O3. *Proceedings of the Institution of Mechanical Engineers, Part C: Journal of Mechanical Engineering Science*, *233*(17), 5933–5944. 10.1177/0954406219860163.

Singh, R., I. Singh, & R. Kumar, Brar G. S. (2019b). Waste Thermosetting Polymer and Ceramic as Reinforcement in Thermoplastic Matrix for Sustainability: Thermomechanical Investigations, *Journal of Thermoplastic composite materials*, *34*(4), 523–535 10.1177/0892705719847237.

10 Hybrid Feedstock Filament Processing for the Preparation of Composite Structures in Heritage Repair

Vinay Kumar, Rupinder Singh, and Inderpreet Singh Ahuja

10.1 INTRODUCTION

In the past two decades, a number of studies have reported on thermoplastic composite materials for innovative product development for various engineering applications; for example, polylactic acid (PLA) matrix composite prepared by reinforcement of almond skin has highlighted novel applications of the composite in scaffolding. Such prototypes may be used for the treatment of fracture sites (Singh et al., 2019a, Kumar et al., 2021a). Researchers have also reported preparation of a biocompatible feedstock filament of fused deposition modeling (FDM) for bioabsorbable prototype fabrication by which low-cost biomedical implants may be manufactured on a large scale (Singh & Ranjan 2018, Kumar et al., 2021b). Like biocompatible hybrid filaments of PLA-chitosan–hydroxiapetite composite, feedstock filaments of acrylonitrile butadiene styrene (ABS), polyamide-6 composite also have been explored by some researchers for hybrid 3D-printing applications (Singh et al., 2016, Singh et al., 2018, Kumar et al., 2021c). The investigations were performed on wear analysis and recycling of high-density polyethylene (HDPE), ABS, polyethylene, and PVDF thermoplastics to prepare the composite by reinforcements like polypyrole, carbon nanotubes, and Mn-doped ZnO (Boparai et al., 2016a, Singh et al., 2019b, Kumar et al., 2020a, Kumar et al., 2021d). The properties obtained in such composites also outlined the usefulness of recycled plastic waste for rapid tooling, heritage repair, and sensor applications (Kumar et al., 2020b, Kumar et al., 2020c). Some studies have highlighted the correlation of rheological, thermal, mechanical, and morphological properties for polymer matrix composites, ABS-graphene and PVDF-graphene, that resulted in better 3D printing of smart functional prototypes (Boparai et al., 2016b, Ranjan et al., 2019, Kumar et al., 2021e). Based on the reported literature, Table 10.1 shows the detailed review

DOI: 10.1201/9781003184164-10

TABLE 10.1
List of various properties/keywords investigated for 3D-printing hybrid filament composites of thermoplastic materials

ID	Term	Occurrences	Relevance Score
1	3D printing processes	5	0.7515
2	3D printing of ABS	3	0.7369
3	Addition polymerization study	9	0.4366
4	Additive manufacturing	6	0.5899
5	Artefacts fabrication	3	1.3842
6	Blend of polymers	4	0.5216
7	Combination of engineering plastic	5	0.9532
8	Comparison of 3D-printing processes	5	0.6128
9	Complex design by 3D printing	3	0.9121
10	Composite filament fabrication	7	0.5648
11	Crystallanity of hybrid composites	4	0.5762
12	De-binding of molecules	4	2.2253
13	Thermal properties	5	1.0772
14	Deposition rate analysis	4	1.0878
15	Deposition modeling technique	7	0.4067
16	Development of customizable products	13	0.5212
17	Diameter of feedstock filament	5	0.3716
18	Differential scanning calorimetry	4	1.0173
19	Dimensional printing of prototypes	3	2.1522
20	DMA analysis of composites	3	2.1342
21	Dosage form for medical use	6	1.9114
22	Drug delivery by 3D printing	7	1.7649
23	DSC analysis	3	1.2509
24	Dynamic mechanical analysis	3	2.1342
25	Electron microscopy of hybrid composites	6	0.9498
26	Elongation behavior of composites	4	1.0952
27	End use of hybrid composites	3	0.7705
28	Ethylene vinyl acetate as solvent	3	1.6996
29	FDM techniques	13	0.406
30	FDM-based 3D printing	3	0.9887
31	Feasibility of polymer composite for FDM	3	0.8269
32	FFF-based polymer processing	3	1.7376
33	Filament feedstock fabrication	4	0.3574
34	Formulation of experimental studies	4	0.6902
35	Fused filament fabrication of smart composites	3	1.3897
36	Hardness of printed produts	3	0.7835
37	HDPE 3D-printing applications	3	0.86
38	Hybrid 3D-printed tiles	3	1.8797
39	Hot melt extrusion of polymers	6	1.9734
40	Improvement in properties of recycled polymer	3	0.9497

TABLE 10.1 (Continued)
List of various properties/keywords investigated for 3D-printing hybrid filament composites of thermoplastic materials

ID	Term	Occurrences	Relevance Score
41	Investigation of rheological properties	8	0.4504
42	Literature survey	3	0.7343
43	Lot/Batch production	3	0.6395
44	Manufacturing of consumer products	5	0.5911
45	Morphological property	3	1.2926
46	Strength (MPa) analysis	4	0.717
47	Nano-composite 3D printing	4	1.7753
48	Optimization of filament fabrication	3	0.3748
49	Part production/prototyping	6	0.3744
50	Part fabrication	12	0.5093
51	Patient studies for drug delivery	4	1.5355
52	Percentage crystallinity	4	1.0433
53	Phase transition during thermal effect	4	0.6203
54	PLA 3D-printing applications	9	0.7156
55	Plastic for engineering products	5	0.4663
56	PVDF and its composites	5	0.6692
57	Polylactic acid bio printing	6	1.0132
58	Polypropylene 3D printing	3	1.1038
59	Possibility for AM applications	3	1.0666
60	Preparation of thermoplastic composites	6	0.989
61	Presence of ceramics in polymers	3	1.1084
62	Printing of conventional composites	5	0.6189
63	Printability of hybrid filaments	6	1.2923
64	Printed part for structures maintenance	3	1.1104
65	Product for heritage repair	7	0.4302
66	Production of customizable products	6	0.7423
67	Reinforcement as a smart element	4	0.6909
68	Review of heritage repair and maintenance	6	0.7591
69	Rheological analysis of hybrid materials	3	1.0704
70	Sample preparation	10	0.6509
71	SEM-based morphological studies	5	1.3155
72	Sintering of reinforced elements	4	2.2253
73	Specimen testing and analysis	11	0.4528
74	State of functional parts in actual work space	3	1.3842
75	Stearic acid for blending plastics	3	1.0702
76	Steps for polymer composite prototyping	5	0.714
77	Tablet 3D/4D printing	4	2.1623
78	Temperature versus time study	10	0.5666
79	Tensile property studies	6	0.4995

(Continued)

TABLE 10.1 (Continued)
List of various properties/keywords investigated for 3D-printing hybrid filament composites of thermoplastic materials

ID	Term	Occurrences	Relevance Score
80	Tensile strength of products	8	0.4214
81	TGA of hybrid prototypes	3	0.8488
82	Thermal stability of 3D parts	7	0.6599
83	Thermo-gravimetric analysis	3	1.0193
84	Use of smart composites	7	0.7306
85	Volumetric study for material consumption	3	2.3247

of various terms and properties investigated by researchers around the globe for 3D-printing hybrid feedstock filaments of different polymer-based composites according to the Web of Science database (for the past 20 years).

Based on Table 10.1, it has been observed that the studies performed on hybrid feedstock filament fabricatrion for various 3D/4D-printing applications resulted in the development of smart 3D prototypes that possess magnetic, bioabsorbable, chemical-based one-way programming, and electro-active properties for 4D applications (Singh et al., 2018, Kumar et al., 2021f, Sharma et al., 2021, Kumar et al., 2021g). Based upon Table 10.1, Figure 10.1 shows the Web of Science data-based web cluster of various investigated proprties of polymers and their composites for hybrid 3D/4D-printing applications.

10.2 RESEARCH GAP AND PROBLEM FORMULATION

The reported literature reveals that numerous studies have been reported on hybrid feedstock filament for various engineering and biomedical applications (like smart tiles and biocompatible implants, etc.), but hitherto little has been reported on the development and use of hybrid and smart feedstock filaments for the development of composites that may be 3D printed and used as customizable solutions for repair of non-structural cracks present in heritage structures. Figure 10.2 highlights the gap in research regarding investigation of smart and hybrid feedstock filament and its 3D printing to develop a hybrid solution for heritage repair work.

The investigations on electrically conducting and non-conducting polymer composite matrices i.e. PVDF-graphene-MnZnO (PGM) and PVDF-$CaCO_3$ (PCC), respectively outlined that hybrid feedstock filament of the same matrix may be considered useful to address the issue of heritage repair. The present work is as an extension of previous studies reported to ascertain the most suitable material processing conditions and desired rheological, mechanical, thermal, and morphological properties of the proposed compositions/proportions for 3D printing a hybrid, smart/customized product to repair the cracks of heritage

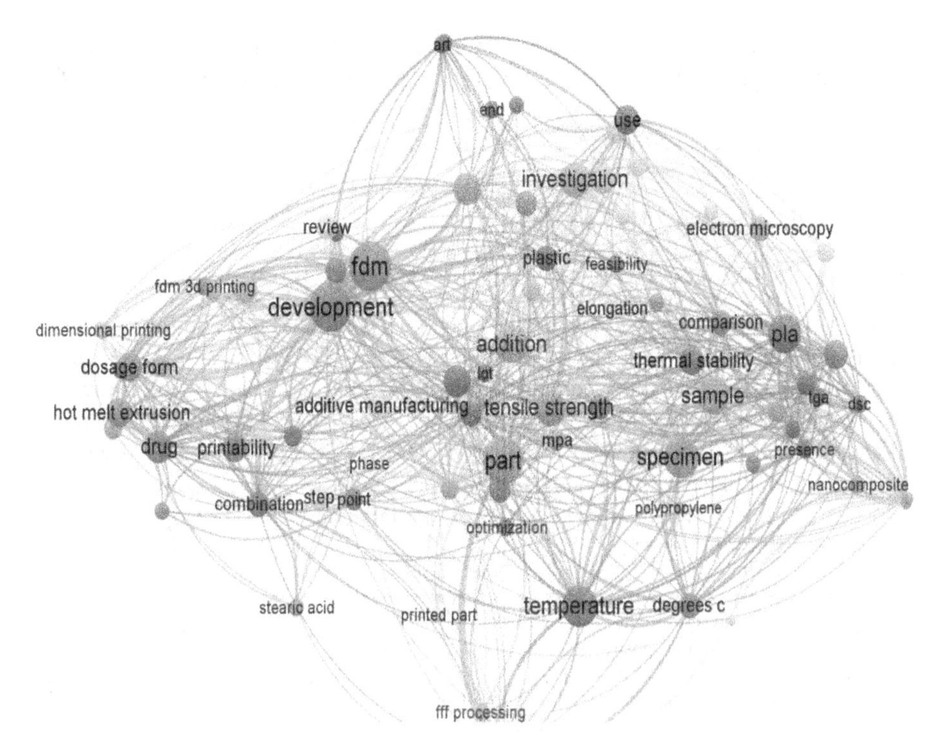

FIGURE 10.1 Web clusters of polymers and their composites for hybrid 3D/4D-printing applications.

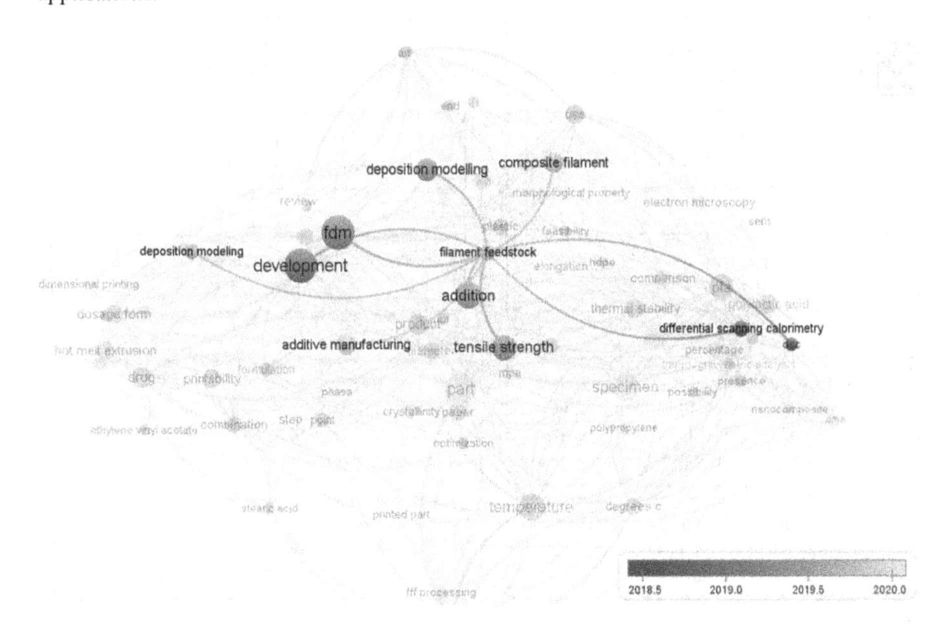

FIGURE 10.2 Research gap highlighted in terms of feedstock filament fabrication for hybridized 3D/4D printing.

structures (Kumar et al., 2021f, g). Therefore, the investigations performed on multi-factor optimization of conducting (PGM) and non-conducting (PCC) composites may support and ascertain the best setting to prepare hybrid feedstock filament for the proposed repair work of heritage structures.

10.3 EXPERIMENTATION

10.3.1 Preparation and Characterization of PVDF Composites

To prepare the PVDF thermoplastic-based smart and hybrid composites with conducting and non-conducting properties, PGM and PCC composites were prepared and characterized for acceptable rheological, thermal, mechanical, morphological, and 4D properties in terms of melt flow index (MFI), differential scanning calorimetry (DSC), universal tensile testing (UTT), porosity, shore-D hardness (SD_H), and stimulus, respectively. Figure 10.3 shows the methodology for the proposed work to prepare and characterize PVDF compositions/proportions for hybrid feedstock filament fabrication.

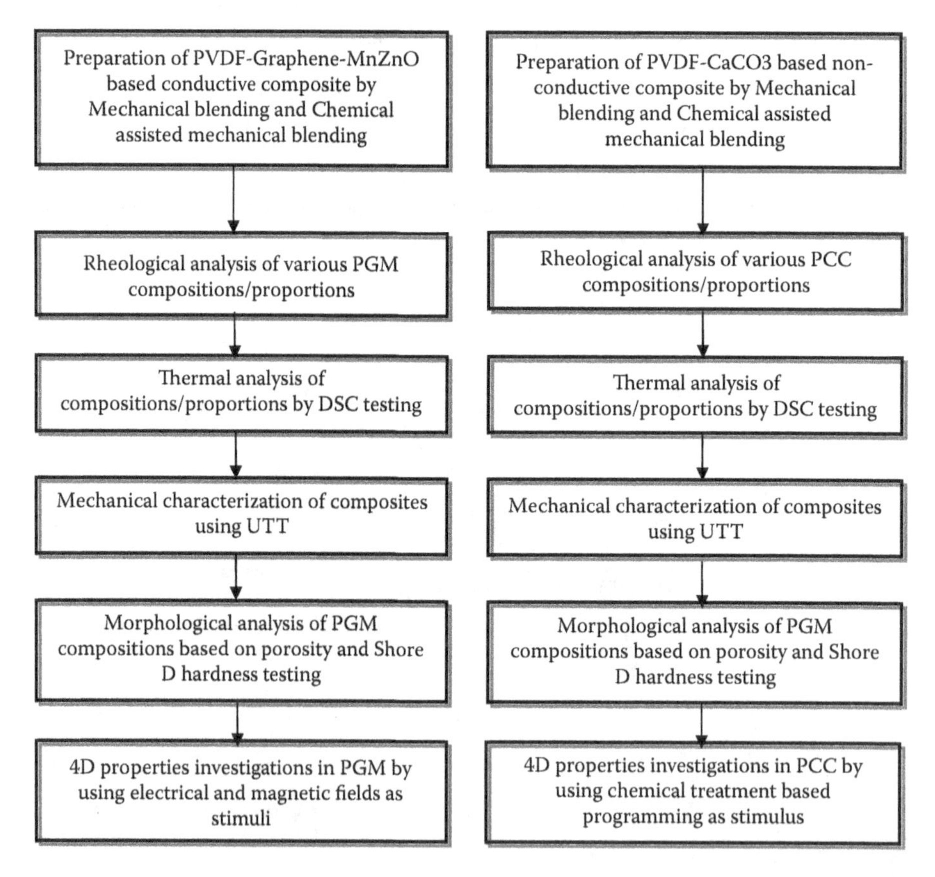

FIGURE 10.3 Work methodologies to prepare and characterize the hybrid filament of PVDF.

From a rheological analysis, it was ascertained that for a conducting-type composite matrix, PVDF-6%graphene-3%MnZnO shows acceptable MFI for 3D printing when prepared by chemical-assisted mechanical blending (CB). Similarly, for a non-conducting-type composition/proportion, PVDF-6%$CaCO_3$ prepared by CB shows the acceptable MFI results. It was observed that the CB route is responsible for better flowability, density, and viscosity characteristics of the PGM and PCC composites due to which the CB process was followed to prepare both compositions/proportions of PVDF (Kumar et al., 2021f, g).

10.3.2 Feedstock Filament Preparation

The CB of PGM and PCC composites was performed on a twin-screw extruder (TSE) to prepare the feedstock filament. Three process parameters with three levels were selected for filament fabrication. Table 10.2 shows the selected input parameters for the TSE while processing both kinds of PVDF composites.

Based on Table 10.2, the Taguchi L9 orthogonal array was used to obtain the nine different settings at which composite feedstock filament can be prepared. This design of the experiment was useful to ascertain the best processing conditions of the TSE for PVDF composites while preparing hybrid filaments for 3D printing. Table 10.3 shows the design of the experiment as per the Taguchi L9 orthogonal array.

TABLE 10.2
Selected input parameters and their levels for PVDF compositions

S. No.	Temperature of Extrusion (°C)	Screw Speed (rpm)	Load (kg)
1	195	75	5
2	205	90	7.5
3	215	105	10

TABLE 10.3
Design of experiment (DOE) for fabrication of PVDF composites by using TSE

S. No.	Temperature of Extrusion (°C)	Screw Speed (rpm)	Load (kg)
1	195	75	5
2	195	90	7.5
3	195	105	10
4	205	75	7.5
5	205	90	10
6	205	105	5
7	215	75	10
8	215	90	5
9	215	105	7.5

The nine wire-shaped filament samples of each composition/proportion were tested for thermal and mechanical properties. Table 10.4 shows the mechanical and morphological properties of the PGM composite in terms of peak strength (PS), SD_H, porosity %, and surface roughness (R_a), whereas Table 10.5 shows the same results obtained for the PCC composite.

It was observed that an extrusion temperature of 215°C, screw speed of 105 rpm, and load of 7.5 kg may be considered the most suitable TSE processing conditions to prepare the hybrid feedstock filament of PGM and PCC composites. This is due to the reason that the best acceptable and desired properties were obtained at these settings. Based on these observations, non-structural cracks were recorded in a heritage building of Grade III (Kumar et al., 2021f, g), shown in Figure 10.4 to prepare a 3D-customized strip of PGM and PCC composite by performing multi-material 3D printing.

TABLE 10.4
Results obtained for PS, SD_H, porosity %, and R_a of PGM composite

S. No.	Average PS (N/mm^2)	Average SD_H (HD)	Average Porosity (%)	Average R_a (μm)
1	28.08 ± 0.05	40.1 ± 0.05	14.35 ± 0.06	110.1 ± 0.05
2	21.52 ± 0.07	38.4 ± 0.06	13.66 ± 0.05	109.8 ± 0.06
3	34.33 ± 0.06	43.7 ± 0.04	12.85 ± 0.06	108.7 ± 0.04
4	38.54 ± 0.04	48.5 ± 0.06	12.32 ± 0.05	107.3 ± 0.05
5	39.31 ± 0.05	43.2 ± 0.05	11.65 ± 0.04	106.5 ± 0.06
6	43.52 ± 0.07	48.4 ± 0.04	10.30 ± 0.05	105.5 ± 0.05
7	43.65 ± 0.05	45.6 ± 0.05	9.52 ± 0.05	103.5 ± 0.04
8	42.40 ± 0.04	47.8 ± 0.07	8.74 ± 0.04	102.3 ± 0.05
9	44.48 ± 0.06	52.9 ± 0.05	8.19 ± 0.05	101.6 ± 0.06

TABLE 10.5
Results obtained for PS, SD_H, porosity %, and R_a of PCC composite

S. No.	Average PS (N/mm^2)	Average ShDH (HD)	Average Porosity (%)	Average R_a (μm)
1	25.27 ± 0.05	39.2 ± 0.04	14.92 ± 0.05	120.3 ± 0.04
2	13.48 ± 0.09	37.5 ± 0.05	13.79 ± 0.03	118.4 ± 0.05
3	32.55 ± 0.04	42.6 ± 0.03	13.59 ± 0.04	117.6 ± 0.07
4	37.70 ± 0.05	47.4 ± 0.05	12. 65 ± 0.07	116.7 ± 0.04
5	38.29 ± 0.06	41.3 ± 0.04	11.77 ± 0.05	116.3 ± 0.05
6	42.78 ± 0.05	47.5 ± 0.05	10.68 ± 0.07	115.5 ± 0.06
7	41.85 ± 0.07	43.7 ± 0.04	10.33 ± 0.06	113.8 ± 0.05
8	41.39 ± 0.06	46.8 ± 0.05	9.52 ± 0.05	112.4 ± 0.03
9	43.41 ± 0.05	51.9 ± 0.06	9.31 ± 0.07	110.1 ± 0.04

FIGURE 10.4 Heritage structure with non-structural cracks on the front wall.

The multi-material FDM-printed customized strip with conductive (PGM) and non-conductive (PCC) layers may be used to repair the wall cracks by self-contracting and self-expanding properties just by providing an external stimulus to the 3D-printed prototype. The stimulus for the customized strip may be an external electric or magnetic field (in the case of PGM composite) and chemical exposure (in the case of PCC) composite. Finally, multi-factor optimization was performed using the Minitab 2019 software package to obtain the optimum composite desirability for fabrication of the hybrid 3D-printer filament and prepare a multi-material composite structure as a repair solution of a crack.

10.4 RESULTS AND DISCUSSION

The results obtained for PGM and PCC composite samples in Table 10.4 and Table 10.5 when processed in the MiniTab software for multi-factor optimization highlighted the efficiency of the proposed model for the preparation of smart and hybridized feedstock filament for 3D-printing applications. The composite desirability of 0.997 was observed for a CB PGM filament fabrication and a 0.993 composite desirability was obtained for a CB PCC composite. Figure 10.5 and Figure 10.6 show the results obtained for multi-factor optimization of a CB PGM and PCC composite, respectively, for the fabrication of a hybrid 3D-printable filament for heritage repair.

These results outlined that a 215°C processing temperature, 105 rpm screw speed, and 7.5 kg applied load are the best settings to achieve acceptable properties in the hybrid filament of PVDF composites for 3D/4D-printing applications. The suggested settings are also in line with the results obtained in Table 10.4 and 10.5, which show that the proposed model is acceptable to prepare a hybrid filament of investigated compositions/proportions to a fabricated multi-material composite structure and repair non-structure wall cracks of heritage structures.

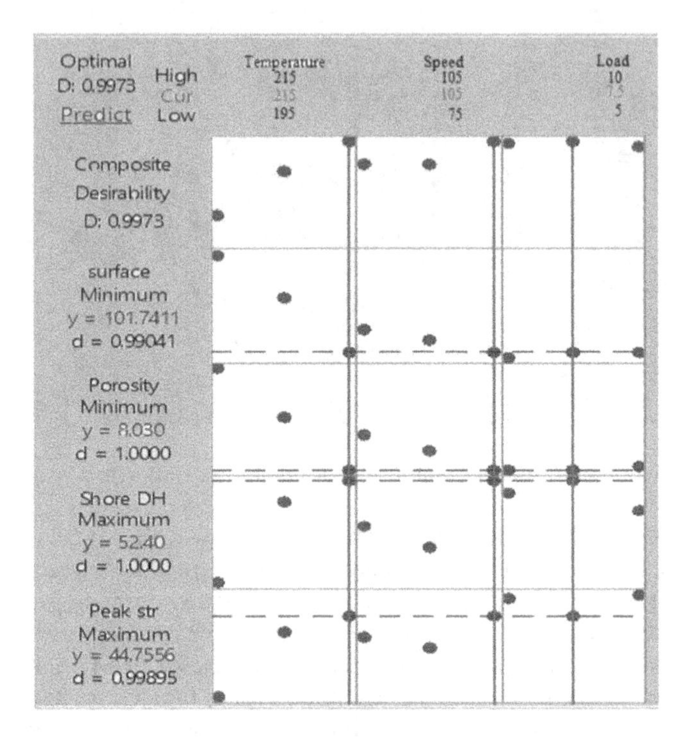

FIGURE 10.5 Multi-factor optimization for CB PVDF-graphene-MnZnO composite filament fabrication.

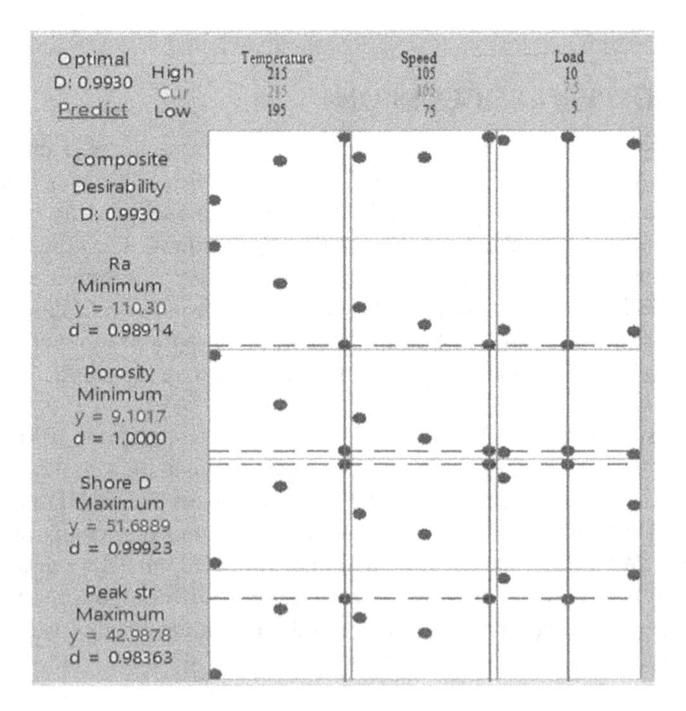

FIGURE 10.6 Multi-factor optimization for CB PVDF-$CaCO_3$ composite filament fabrication.

10.5 SUMMARY

This study highlighted the use of hybrid feedstock filament for FDM-based engineering and biomedical applications with a gap in literature regarding the development and use of smart polymer matrix-based hybrid filament for repair and maintenance of heritage structures. The investigations performed on rheological, thermal, mechanical, morphological, and 4D properties of PVDF-6%graphene-3% MnZnO (conductive) and PVDF-6%$CaCO_3$ (non-conductive) composites outlined that the proposed composites may be considered for preparation of 3D-printer hybrid filament to repair cracks of heritage structures.

The filament fabrication of PVDF composites (prepared by CB process) using TSE and multi-factor optimization of the results obtained highlighted that a 215°C processing temperature, 105 rpm screw speed, and 7.5 kg applied load are the acceptable process parametric conditions to prepare a hybrid filament.

ACKNOWLEDGEMENT

The authors are grateful to the Department of Science and Technology (DST)-Government of India for providing the financial support for this study under the DST-SHRI Project (File No: DST/TDT/SHRI-35/2018).

REFERENCES

Boparai K. S., Singh R., Singh H. (2016a Mar 21). Wear behavior of FDM parts fabricated by composite material feed stock filament. *Rapid Prototyping Journal*, *22*(2), 350–357.

Boparai K. S., Singh R., Singh H. (2016b Mar 21). Experimental investigations for development of Nylon6-Al-Al_2O_3 alternative FDM filament. *Rapid Prototyping Journal*, *22*(2), 217–224.

Kumar V., Singh R., Ahuja I. P. (2020a May 18). Secondary recycled acrylonitrile–butadiene–styrene and graphene composite for 3D/4D applications: rheological, thermal, magnetometric, and mechanical analyses. *Journal of Thermoplastic Composite Materials*, 0892705720925114.

Kumar R., Pandey A. K., Singh R., Kumar V. (2020b Dec). On nano polypyrrole and carbon nano tube reinforced PVDF for 3D printing applications: Rheological, thermal, electrical, mechanical, morphological characterization. *Journal of Composite Materials*, *54*(29), 4677–4689.

Kumar V., Singh R., Ahuja I. P., Hashmi M. J. (2020c Jan 2). On technological solutions for repair and rehabilitation of heritage sites: a review. *Advances in Materials and Processing Technologies*, *6*(1), 146–166.

Kumar R., Singh R., Kumar V., Kumar P., Prakesh C., Singh S. (2021a Apr 26). Characterization of in-House-Developed Mn-ZnO-Reinforced Polyethylene: A Sustainable Approach for Developing Fused Filament Fabrication-Based Filament. *Journal of Materials Engineering and Performance*, *30*, 5368–5382.

Kumar R., Singh R., Kumar V., Kumar P. (2021b Feb 1). On Mn doped ZnO nano particles reinforced in PVDF matrix for fused filament fabrication: Mechanical, thermal, morphological and 4D properties. *Journal of Manufacturing Processes*, *62*, 817–832.

Kumar V., Singh R., Ahuja I. P. (2021c Jun 9). On 4D capabilities of chemical assisted mechanical blended ABS-nano graphene composite matrix. *Materials Today: Proceedings*. 10.1016/j.matpr.2021.05.678.

Kumar, V., Singh, R. and Ahuja, I. P. S. (2021d). On Correlation of Rheological, Thermal, Mechanical and Morphological Properties of Mechanically Blended PVDF-Graphene Composite for 4d Applications. Elsevier.

Kumar V., Singh R., Ahuja I. P. S. *et al.* (2021e). On Nanographene-Reinforced Polyvinylidene Fluoride Composite Matrix for 4D Applications. *Journal of Materials Engineering and Performance*, *30*, 4860–4871. 10.1007/s11665-021-05459-z

Kumar V., Singh R., and Ahuja I. P. S. (2021f). On programming of PVDF-$CaCO_3$ composite for 4D printing applications in heritage structures. *Proceedings of the Institution of Mechanical Engineers, Part L: Journal of Materials: Design and Applications*. In Press. 10.1177/14644207211044298.

Kumar V., Singh R., Ahuja I. P. (2021g Apr 26). On correlation of rheological, thermal, mechanical and morphological properties of chemical assisted mechanically blended ABS-Graphene composite as tertiary recycling for 3D printing applications.*Advances in Materials and Processing Technologies*, 1–20. 10.1080/2374068X.2021.1913324.

Ranjan N., Singh R., Ahuja I. S., Singh J. (2019). Fabrication of PLA-HAp-CS based biocompatible and biodegradable feedstock filament using twin screw extrusion. In *Additive manufacturing of emerging materials* (pp. 325–345). Springer, Cham.

Sharma R., Singh R., Batish A., Ranjan N. (2021 Sep 27). On synergistic effect of BaTiO3 and graphene reinforcement in polyvinyl diene fluoride matrix for four dimensional applications. *Proceedings of the Institution of Mechanical Engineers, Part C: Journal of Mechanical Engineering Science*, 10.1177/09544062211015763.

Singh R., Kumar R., Ahuja I. P. (2018 Nov 12). Mechanical, thermal and melt flow of aluminum-reinforced PA6/ABS blend feedstock filament for fused deposition modeling. *Rapid Prototyping Journal*, *24*(9), 1455–1468.

Singh R., Kumar R., Pawanpreet, Singh M, Singh J. (2019a Nov 5). On mechanical, thermal and morphological investigations of almond skin powder-reinforced polylactic acid feedstock filament. *Journal of Thermoplastic Composite Materials*, 10.1177/0892705719886010.

Singh R., Kumar R., Tiwari S., Vishwakarma S., Kakkar S., Rajora V., Bhatoa S. (2019b Jul 22). On secondary recycling of ZrO2-reinforced HDPE filament prepared from domestic waste for possible 3-D printing of bearings. *Journal of Thermoplastic Composite Materials*. 10.1177/0892705719864628.

Singh R., Ranjan N. (2018 Nov). Experimental investigations for preparation of biocompatible feedstock filament of fused deposition modeling (FDM) using twin screw extrusion process. *Journal of Thermoplastic Composite Materials*, *31*(11), 1455–1469.

Singh R., Singh S., Mankotia K. (2016 Mar 21). Development of ABS based wire as feedstock filament of FDM for industrial applications. *Rapid Prototyping Journal*, *22*(2), 300–310.

Index